London Mathematical Society Student Texts 28

Potential Theory in the Complex Plane

Thomas Ransford
Université Laval, Quebec

CAMBRIDGE
UNIVERSITY PRESS

Published by the Press Syndicate of the University of Cambridge
The Pitt Building, Trumpington Street, Cambridge CB2 1RP
40 West 20th Street, New York, NY 10011-4211, USA
10 Stamford Road, Oakleigh, Melbourne 3166, Australia

© Cambridge University Press 1995

First published 1995

Printed in Great Britain at the University Press, Cambridge

A catalogue record for this book is available from the British Library

Library of Congress cataloging in publication data
Ransford, Thomas
Potential theory in the complex plane / Thomas Ransford
  p. cm. - (London Mathematical Society student texts; 28)
Includes bibliographical references and index
ISBN 0521 46120 0. - ISBN 0521 46654 7 (pbk.)
1. Potential theory (Mathematics) 2. Functions of complex variables
I. Title. II. Series.
QA404.7.R36 1995
515.9-dc20   94-38846 CIP

ISBN 0 521 46120 0 hardback
ISBN 0 521 46654 7 paperback

To Line

Contents

Preface   ix

A Word about Notation   1

1   Harmonic Functions   **3**
   1.1   Harmonic and Holomorphic Functions   3
   1.2   The Dirichlet Problem on the Disc   7
   1.3   Positive Harmonic Functions   13
   Notes on Chapter 1.   22

2   Subharmonic Functions   **25**
   2.1   Upper Semicontinuous Functions   25
   2.2   Subharmonic Functions   27
   2.3   The Maximum Principle   29
   2.4   Criteria for Subharmonicity   35
   2.5   Integrability   39
   2.6   Convexity   43
   2.7   Smoothing   48
   Notes on Chapter 2.   51

3   Potential Theory   **53**
   3.1   Potentials   53
   3.2   Polar Sets   55
   3.3   Equilibrium Measures   58
   3.4   Upper Semicontinuous Regularization   62
   3.5   Minus-Infinity Sets   65
   3.6   Removable Singularities   67
   3.7   The Generalized Laplacian   71
   3.8   Thinness   78
   Notes on Chapter 3.   82

4 The Dirichlet Problem — 85
- 4.1 Solution of the Dirichlet Problem — 85
- 4.2 Criteria for Regularity — 91
- 4.3 Harmonic Measure — 96
- 4.4 Green's Functions — 106
- 4.5 The Poisson–Jensen Formula — 116
- Notes on Chapter 4 — 126

5 Capacity — 127
- 5.1 Capacity as a Set Function — 127
- 5.2 Computation of Capacity — 132
- 5.3 Estimation of Capacity — 137
- 5.4 Criteria for Thinness — 146
- 5.5 Transfinite Diameter — 152
- Notes on Chapter 5 — 160

6 Applications — 161
- 6.1 Interpolation of L^p-spaces — 161
- 6.2 Homogeneous Polynomials — 166
- 6.3 Uniform Approximation — 169
- 6.4 Banach Algebras — 176
- 6.5 Complex Dynamics — 190
- Notes on Chapter 6 — 206

A Borel Measures — 209
- A.1 Supports — 209
- A.2 Regularity — 210
- A.3 Radon Measures — 211
- A.4 Weak*-Convergence — 216

Bibliography — 219

Index — 225

Glossary of Notation — 231

Preface

When first learning potential theory, as a new graduate student, I experienced some difficulty with the literature then available. The choice lay between several excellent but encyclopaedic treatises on the subject, from which it was hard work to extract what was needed, and several equally excellent books on complex variable, each containing a useful chapter on potential theory, but which did not go nearly far enough. This book is an attempt to bridge that gap—indeed it was consciously written as the book that I should have liked to read all those years ago.

Potential theory is the name given to the broad field of analysis encompassing such topics as harmonic and subharmonic functions, the Dirichlet problem, harmonic measure, Green's functions, potentials and capacity. It can be developed in many contexts, ranging from classical potential theory in $\mathbb{R}^n$ and pluripotential theory in $\mathbb{C}^n$ to axiomatic theories in very general spaces. In between there are versions relating to Riemann surfaces and other manifolds, uniform algebras and analytic multifunctions, to say nothing of the connections with Brownian motion and other stochastic processes. However, there is one case which is common to them all: potential theory in the plane. As it contains all the essential ingredients of the subject, yet is relatively easy and quick to treat, it seems to me to be well worth mastering first. This is the subject of the book.

There is also a further goal, hinted at by the use of the word 'complex' in the title. It is to emphasize the very close connection between potential theory and complex analysis. This works both ways. In one direction, the techniques of complex analysis, particularly conformal mapping, can be used to speed up and simplify proofs of some of the results in potential theory. Going the other way, these same theorems in potential theory have a multitude of applications in complex analysis. Examples of the latter are scattered liberally throughout the text, including, for example: Picard's theorem, the Phragmén–Lindelöf principle, the Radó–Stout theorem, Lindelöf's theorem on asymptotic values, the Riemann mapping theorem (including continuity at the boundary), the Koebe one-quarter theorem, Hilbert's lemniscate theorem, and the sharp quantitative form of Runge's theorem.

Chapters 1–5 cover the basic theory. They are pitched at a level suitable for a first-year graduate student. Thus they presuppose a knowledge of elementary complex analysis (Cauchy's theorem, Cauchy's integral formula, the maximum modulus principle, the identity principle and Taylor's theorem), and also of basic measure theory (Lebesgue measure, the monotone convergence theorem, the dominated convergence theorem, Fatou's lemma and Fubini's theorem). On the other hand, it is not assumed that the reader has previously encountered general Borel measures, so there is an appendix on these, including a proof of the Riesz representation theorem. There are exercises at the end of each section, ranging from five-finger problems to thinly-disguised new theorems.

Chapter 6 demands rather more of the reader. It contains a variety of applications of potential theory to other areas of analysis, notably to functional analysis, approximation theory and dynamical systems. Some, like the results on L^p-interpolation and uniform approximation, are classical, while others, such as those relating to spectral theory of Banach algebras and to Hausdorff dimension of Julia sets, are relatively recent. My own interest in potential theory was inspired by such applications, and I have tried to pick a selection that conveys some of the attraction that the subject holds for me. I hope that it also shows that, even though the basic theory is by now fifty years old, there are still interesting new applications to be found.

This book is based upon versions of a graduate course that I gave at Leeds, Cambridge and Brown Universities, and I should like to thank all those who attended, students and faculty, for their invaluable questions and comments. I owe a particular debt of gratitude to Andy Browder, Brian Cole and John Wermer, who encouraged me to write the course up. Also I am grateful to David Tranah and Roger Astley of Cambridge University Press, as well as an anonymous referee, whose suggestions helped shape the book. And finally, I wish to say a big thank-you to my wife Line: for tolerating all my moaning and groaning about the text, for lending me her slow but trusty Mac SE on which most of it was typed, and for producing a new 'theorem', Julian Vincent, just before its completion. I dedicate the book to her.

Cap Rouge, Québec
June 1994

T.J.R.

A Word about Notation

Much of the action in this book takes on a domain (i.e. a non-empty, connected, open set) in the complex plane $\mathbb{C}$ or the Riemann sphere $\mathbb{C}_\infty$. Though working in the latter may seem out of keeping with the title of the book, it is often beneficial to do so even if one's primary interest is in plane domains, not least because closed subsets of the sphere are always compact. For this reason we shall adopt the convention that, given a subset S of $\mathbb{C}$ (or $\mathbb{C}_\infty$), its closure $\overline{S}$ and its boundary ∂S will *always* be taken relative to $\mathbb{C}_\infty$. In particular, it is important to remember that if S is an unbounded subset of $\mathbb{C}$ then $\infty \in \partial S$; reminders about this will be issued from time to time.

The other piece of notation that will be needed straightaway is a symbol for discs. Given $w \in \mathbb{C}$ and $\rho > 0$, we write

$$\Delta(w, \rho) := \{z \in \mathbb{C} : |z - w| < \rho\},$$
$$\overline{\Delta}(w, \rho) := \{z \in \mathbb{C} : |z - w| \le \rho\}.$$

Although Δ is also the standard symbol for the Laplacian, namely

$$\Delta f := f_{xx} + f_{yy},$$

no confusion should arise since one Δ is followed by a bracket and the other by a function!

All the remaining notation will be introduced as it is needed, and summarized in a glossary at the end of the book. We are now ready to begin.

Chapter 1

Harmonic Functions

1.1 Harmonic and Holomorphic Functions

Harmonic functions, namely solutions of Laplace's equation, exhibit many properties reminiscent of those of holomorphic functions. In fact, when working in the plane, as we shall, there is a direct connection between the two classes. We shall unashamedly exploit this to accelerate the initial development of harmonic functions, under the assumption that we already know something about holomorphic ones. Later, potential theory will repay its debt to complex analysis in the form of many beautiful applications.

We begin with the formal definition.

Definition 1.1.1 Let U be an open subset of $\mathbb{C}$. A function $h\colon U \to \mathbb{R}$ is called *harmonic* if $h \in C^2(U)$ and $\Delta h = 0$ on U.

The following basic result not only furnishes numerous examples of harmonic functions, but also provides a useful tool in deriving their elementary properties from those of holomorphic functions.

Theorem 1.1.2 *Let D be a domain in $\mathbb{C}$.*

(a) *If f is holomorphic on D and $h = \operatorname{Re} f$, then h is harmonic on D.*

(b) *If h is harmonic on D, and if D is simply connected, then $h = \operatorname{Re} f$ for some f holomorphic on D. Moreover f is unique up to adding a constant.*

Proof. (a) Writing $f = h + ik$, the Cauchy–Riemann equations give that $h_x = k_y$ and $h_y = -k_x$. Therefore

$$\Delta h = h_{xx} + h_{yy} = k_{yx} - k_{xy} = 0.$$

3

(b) If $h = \operatorname{Re} f$ for some holomorphic function f, say $f = h + ik$, then

(1.1) $$f' = h_x + ik_x = h_x - ih_y.$$

Thus, if f exists, then f' is completely determined by h, and hence f is unique up to adding a constant.

Equation (1.1) also suggests how we might construct such a function f. Define $g: D \to \mathbb{C}$ by
$$g = h_x - ih_y.$$
Then $g \in C^1(D)$ and g satisfies the Cauchy–Riemann equations because
$$h_{xx} = -h_{yy} \quad \text{and} \quad h_{xy} = h_{yx}.$$
Therefore g is holomorphic on D. Fix $z_0 \in D$, and define $f: D \to \mathbb{C}$ by
$$f(z) = h(z_0) + \int_{z_0}^{z} g(w)\,dw,$$
the integral being taken over any path in D from z_0 to z. As D is simply connected, Cauchy's theorem ensures that the integral is independent of the particular path chosen. Then f is holomorphic on D and $f' = g = h_x - ih_y$. Writing $\tilde{h} = \operatorname{Re} f$, we have
$$\tilde{h}_x - i\tilde{h}_y = f' = h_x - ih_y,$$
so that $(\tilde{h} - h)_x \equiv 0$ and $(\tilde{h} - h)_y \equiv 0$. It follows that $\tilde{h} - h$ is constant on D, and putting $z = z_0$ shows that the constant is zero. Thus indeed $h = \operatorname{Re} f$. □

As a consequence, we obtain a useful result about holomorphic logarithms.

Corollary 1.1.3 *Let f be holomorphic and non-zero on a simply connected domain D in $\mathbb{C}$. Then there exists a holomorphic function g on D such that $f = e^g$.*

Proof. Put $h = \log|f|$ on D. Because h is locally the real part of a holomorphic function, namely a branch of $\log f$, it is harmonic by Theorem 1.1.2 (a). By Theorem 1.1.2 (b) there exists g holomorphic on D such that $h = \operatorname{Re} g$ there, or in other words, $|fe^{-g}| = 1$ on D. By the maximum principle for holomorphic functions, fe^{-g} is a constant C. Adding a suitable constant to g, we can suppose that $C = 1$, and so $f = e^g$. □

1.1. HARMONIC AND HOLOMORPHIC FUNCTIONS

Corollary 1.1.3 (and, by implication, Theorem 1.1.2 (b)) may fail if D is not simply connected. For example, the function $f(z) = z$ is holomorphic and non-zero on the domain $D = \mathbb{C} \setminus \{0\}$; but there is no holomorphic function g such that $z = e^{g(z)}$ on this domain, for such a g would satisfy $g'(z) = 1/z$, and this would then imply that

$$0 = \int_{|z|=1} g'(z)\, dz = \int_{|z|=1} \frac{1}{z}\, dz = 2\pi i,$$

which is obviously false.

However, since discs are simply connected, every harmonic function is at least locally the real part of some holomorphic function. This has the following immediate consequences.

Corollary 1.1.4 *If h is a harmonic function on an open subset U of $\mathbb{C}$, then $h \in C^\infty(U)$.* □

Corollary 1.1.5 *If $f: U_1 \to U_2$ is a holomorphic map between open subsets U_1, U_2 of $\mathbb{C}$, and if h is harmonic on U_2, then $h \circ f$ is harmonic on U_1.* □

This result allows us to extend the notion of harmonicity to the Riemann sphere. Given a function h defined on an open neighbourhood U of ∞, we say h is harmonic on U if $h \circ \phi^{-1}$ is harmonic on $\phi(U)$, where ϕ is a conformal mapping of U onto an open subset of $\mathbb{C}$. It does not matter which map ϕ is chosen: if ϕ_1 and ϕ_2 are two such choices, then $(h \circ \phi_1^{-1}) = (h \circ \phi_2^{-1}) \circ f$, where $f = \phi_2 \circ \phi_1^{-1}$, so by Corollary 1.1.5 $h \circ \phi_1^{-1}$ is harmonic on $\phi_1(U)$ if and only if $h \circ \phi_2^{-1}$ is harmonic on $\phi_2(U)$.

Another simple consequence of Theorem 1.1.2 will be of great importance later.

Theorem 1.1.6 (Mean-Value Property) *Let h be a function harmonic on an open neighbourhood of the disc $\overline{\Delta}(w, \rho)$. Then*

$$h(w) = \frac{1}{2\pi} \int_0^{2\pi} h(w + \rho e^{i\theta})\, d\theta.$$

Proof. Choose $\rho' > \rho$ so that h is harmonic on $\Delta(w, \rho')$. Applying Theorem 1.1.2 (b), there exists f holomorphic on $\Delta(w, \rho')$ such that $h = \operatorname{Re} f$ there. By Cauchy's integral formula, we have

$$f(w) = \frac{1}{2\pi i} \int_{|\zeta - w| = \rho} \frac{f(\zeta)}{\zeta - w}\, d\zeta = \frac{1}{2\pi} \int_0^{2\pi} f(w + \rho e^{i\theta})\, d\theta.$$

The result follows upon taking real parts of both sides. □

This section ends with two further ways in which harmonic functions behave like holomorphic ones, an identity principle and a maximum principle. We shall deduce the harmonic versions of both these results from their holomorphic counterparts.

Theorem 1.1.7 (Identity Principle) *Let h and k be harmonic functions on a domain D in $\mathbb{C}$. If $h = k$ on a non-empty open subset U of D, then $h = k$ throughout D.*

Proof. We can suppose, without loss of generality, that $k = 0$. Set $g = h_x - ih_y$. Then as in the proof of Theorem 1.1.2, g is holomorphic on D, and also $g = 0$ on U since $h = 0$ there. By the identity principle for holomorphic functions, it follows that $g = 0$ throughout D, and hence that $h_x = 0$ and $h_y = 0$ on D. Therefore h is constant on D, and since $h = 0$ on U, this constant must be zero. □

For holomorphic functions, a stronger form of identity principle holds: namely, if two holomorphic functions agree on a set with a limit point in the domain D, then they agree throughout D. However, this is not the case for harmonic functions. For instance, the functions $h(z) = \operatorname{Re} z$ and $k(z) = 0$ are harmonic on $\mathbb{C}$ and agree on the imaginary axis without being equal on the whole of $\mathbb{C}$.

Theorem 1.1.8 (Maximum Principle) *Let h be a harmonic function on a domain D in $\mathbb{C}$.*

(a) *If h attains a local maximum on D, then h is constant.*

(b) *If h extends continuously to $\overline{D}$ and $h \leq 0$ on ∂D, then $h \leq 0$ on D.*

This is perhaps a timely moment for a reminder about our convention that all closures and boundaries are taken with respect to $\mathbb{C}_\infty$ rather than $\mathbb{C}$. Indeed Theorem 1.1.8 (b) would otherwise be false: consider, for example, the harmonic function $h(z) = \operatorname{Re} z$ on the domain $D = \{z \in \mathbb{C} : \operatorname{Re} z > 0\}$.

Proof. (a) Suppose that h attains a local maximum at $w \in D$. Then for some $r > 0$ we have $h \leq h(w)$ on $\Delta(w, r)$. By Theorem 1.1.2 (b) there exists a function f holomorphic on $\Delta(w, r)$ such that $h = \operatorname{Re} f$ there. Then $|e^f|$ attains a local maximum at w, so e^f must be constant. Therefore h is constant on $\Delta(w, r)$, and hence on the whole of D by the identity principle.

(b) As $\overline{D}$ is compact, h must attain a maximum at some point $w \in \overline{D}$. If $w \in \partial D$, then $h(w) \leq 0$ by assumption, and so $h \leq 0$ on D. If $w \in D$, then by part (a) h is constant on D, hence on $\overline{D}$, and so once again $h \leq 0$ on D. □

Exercises 1.1

1. Let $h(x+iy) = e^x(x\cos y - y\sin y)$. Show that h is harmonic on $\mathbb{C}$, and find a holomorphic function f on $\mathbb{C}$ such that $h = \operatorname{Re} f$.

2. Let h be a function harmonic on $\{z \in \mathbb{C} : \rho_1 < |z| < \rho_2\}$. Using the fact that $h_x - ih_y$ is holomorphic, prove that there exist unique constants $(a_n)_{n \in \mathbb{Z}}$ and b, with $a_0, b \in \mathbb{R}$, such that
$$h(z) = \operatorname{Re}\left(\sum_{-\infty}^{\infty} a_n z^n\right) + b\log|z| \quad (\rho_1 < |z| < \rho_2).$$

3. Let h, k be functions which are harmonic and non-constant on a domain D. Prove that hk is harmonic if and only if $h+ick$ is holomorphic for some real constant c. [Hint for the 'only if': consider f/g, where $f = h_x - ih_y$ and $g = k_x - ik_y$.]

4. Show that every harmonic function is real-analytic, and use this to give another proof of the identity principle.

5. Show that the only functions harmonic on the whole of $\mathbb{C}_\infty$ are the constants.

1.2 The Dirichlet Problem on the Disc

The Dirichlet problem is to find a harmonic function on a domain with prescribed boundary values. It is one of the great advantages of harmonic functions over holomorphic ones that for 'nice' domains, a solution always exists. This is a powerful tool with many applications.

Here is the formal statement of the problem.

Definition 1.2.1 Let D be a subdomain of $\mathbb{C}$, and let $\phi\colon \partial D \to \mathbb{R}$ be a continuous function. The *Dirichlet problem* is to find a harmonic function h on D such that $\lim_{z \to \zeta} h(z) = \phi(\zeta)$ for all $\zeta \in \partial D$.

The question of uniqueness is easily settled.

Theorem 1.2.2 (Uniqueness Theorem) *With the notation of Definition 1.2.1, there is at most one solution h to the Dirichlet problem.*

Proof. Suppose that h_1 and h_2 are both solutions. Then $h_1 - h_2$ is harmonic on D, extends continuously to $\overline{D}$, and is zero on ∂D. Applying the maximum principle Theorem 1.1.8 (b) to $\pm(h_1 - h_2)$, we conclude that $h_1 - h_2 = 0$. $\square$

The question of existence of solutions to the Dirichlet problem is rather more delicate and is postponed until Chapter 4. However, there is one important special case that we can solve now, namely when D is a disc. To this end, we make the following definition.

Definition 1.2.3 (a) The *Poisson kernel* $P: \Delta(0,1) \times \partial\Delta(0,1) \to \mathbb{R}$ is defined by

$$P(z,\zeta) := \operatorname{Re}\left(\frac{\zeta+z}{\zeta-z}\right) = \frac{1-|z|^2}{|\zeta-z|^2} \quad (|z|<1, |\zeta|=1).$$

(b) If $\Delta = \Delta(w,\rho)$ and $\phi: \partial\Delta \to \mathbb{R}$ is a Lebesgue-integrable function, then its *Poisson integral* $P_\Delta \phi: \Delta \to \mathbb{R}$ is defined by

$$P_\Delta\phi(z) := \frac{1}{2\pi} \int_0^{2\pi} P\left(\frac{z-w}{\rho}, e^{i\theta}\right) \phi(w+\rho e^{i\theta}) \, d\theta \quad (z \in \Delta).$$

More explicitly, if $r < \rho$ and $0 \le t < 2\pi$, then

$$P_\Delta\phi(w+re^{it}) = \frac{1}{2\pi} \int_0^{2\pi} \frac{\rho^2 - r^2}{\rho^2 - 2\rho r \cos(\theta - t) + r^2} \phi(w + \rho e^{i\theta}) \, d\theta.$$

The following result is fundamental.

Theorem 1.2.4 *With the notation of Definition 1.2.3:*

(a) $P_\Delta \phi$ *is harmonic on* Δ;

(b) *if ϕ is continuous at $\zeta_0 \in \partial\Delta$, then* $\lim_{z \to \zeta_0} P_\Delta\phi(z) = \phi(\zeta_0)$.

In particular, if ϕ is continuous on the whole of $\partial\Delta$, then $h := P_\Delta \phi$ solves the Dirichlet problem on Δ.

Proof. (a) Making an affine change of variable if necessary, we can suppose that $w = 0$ and $\rho = 1$, so that $\Delta = \Delta(0,1)$. Then

$$P_\Delta\phi(z) = \operatorname{Re}\left(\frac{1}{2\pi} \int_0^{2\pi} \frac{e^{i\theta}+z}{e^{i\theta}-z} \phi(e^{i\theta}) \, d\theta\right) \quad (z \in \Delta),$$

so that $P_\Delta \phi$ is the real part of a holomorphic function of z. Hence it is harmonic on Δ.

Turning to the proof of part (b), it is convenient first to prove a lemma about the Poisson kernel.

1.2. THE DIRICHLET PROBLEM ON THE DISC

Lemma 1.2.5 *The Poisson kernel P satisfies:*

(i) $P(z,\zeta) > 0$ $(|z| < 1, |\zeta| = 1)$;

(ii) $\frac{1}{2\pi} \int_0^{2\pi} P(z, e^{i\theta}) \, d\theta = 1$ $(|z| < 1)$;

(iii) $\sup_{|\zeta - \zeta_0| \geq \delta} P(z,\zeta) \to 0$ as $z \to \zeta_0$ $(|\zeta_0| = 1, \delta > 0)$.

Proof. (i) This is clear from the definition of $P(z,\zeta)$.

(ii) Expressing the given integral as a contour integral and using the Cauchy integral formula, we obtain

$$\begin{aligned}
\frac{1}{2\pi} \int_0^{2\pi} P(z, e^{i\theta}) \, d\theta &= \operatorname{Re}\left(\frac{1}{2\pi i} \int_{|\zeta|=1} \frac{\zeta+z}{\zeta-z} \frac{d\zeta}{\zeta}\right) \\
&= \operatorname{Re}\left(\frac{1}{2\pi i} \int_{|\zeta|=1} \left(\frac{2}{\zeta-z} - \frac{1}{\zeta}\right) d\zeta\right) \\
&= \operatorname{Re}(2-1) = 1.
\end{aligned}$$

(iii) If $|z - \zeta_0| < \delta$ then

$$\sup_{|\zeta-\zeta_0|\geq\delta} P(z,\zeta) \leq \frac{1-|z|^2}{(\delta - |\zeta_0 - z|)^2},$$

and the result follows easily from this. $\square$

Proof of Theorem 1.2.4 (b). Once again, we may suppose that $\Delta = \Delta(0,1)$. Then using Lemma 1.2.5 (i) and (ii) we have

$$\begin{aligned}
|P_\Delta \phi(z) - \phi(\zeta_0)| &= \left| \frac{1}{2\pi} \int_0^{2\pi} P(z, e^{i\theta})(\phi(e^{i\theta}) - \phi(\zeta_0)) \, d\theta \right| \\
&\leq \frac{1}{2\pi} \int_0^{2\pi} P(z, e^{i\theta}) |\phi(e^{i\theta}) - \phi(\zeta_0)| \, d\theta.
\end{aligned}$$

Let $\epsilon > 0$. If ϕ is continuous at ζ_0, then there exists $\delta > 0$ such that

$$|\zeta - \zeta_0| < \delta \Rightarrow |\phi(\zeta) - \phi(\zeta_0)| < \epsilon.$$

Hence, using Lemma 1.2.5 (i) and (ii) again, it follows that

$$\frac{1}{2\pi} \int_{|e^{i\theta}-\zeta_0|<\delta} P(z, e^{i\theta})|\phi(e^{i\theta}) - \phi(\zeta_0)| \, d\theta \leq \frac{1}{2\pi} \int_0^{2\pi} P(z, e^{i\theta}) \epsilon \, d\theta = \epsilon.$$

Also, from Lemma 1.2.5 (iii), there exists $\delta' > 0$ such that

$$|z - \zeta_0| < \delta' \Rightarrow \sup_{|\zeta-\zeta_0|\geq\delta} P(z,\zeta) < \epsilon.$$

Hence if $|z - \zeta_0| < \delta'$ then

$$\frac{1}{2\pi} \int_{|e^{i\theta} - \zeta_0| \geq \delta} P(z, e^{i\theta}) |\phi(e^{i\theta}) - \phi(\zeta_0)| \, d\theta$$
$$\leq \frac{1}{2\pi} \int_0^{2\pi} \epsilon |\phi(e^{i\theta}) - \phi(\zeta_0)| \, d\theta$$
$$\leq \epsilon \left(\frac{1}{2\pi} \int_0^{2\pi} |\phi(e^{i\theta})| \, d\theta + |\phi(\zeta_0)| \right).$$

Combining these facts, we deduce that if $|z - \zeta_0| < \delta'$ then

$$|P_\Delta \phi(z) - \phi(\zeta_0)| \leq \epsilon \left(1 + \frac{1}{2\pi} \int_0^{2\pi} |\phi(e^{i\theta})| \, d\theta + |\phi(\zeta_0)| \right).$$

This concludes the proof. □

As an immediate consequence of this result, we obtain an analogue of the Cauchy integral formula for harmonic functions.

Corollary 1.2.6 (Poisson Integral Formula) *If h is harmonic on an open neighbourhood of the disc $\overline{\Delta}(w, \rho)$, then for $r < \rho$ and $0 \leq t < 2\pi$*

$$h(w + re^{it}) = \frac{1}{2\pi} \int_0^{2\pi} \frac{\rho^2 - r^2}{\rho^2 - 2\rho r \cos(\theta - t) + r^2} h(w + \rho e^{i\theta}) \, d\theta.$$

Proof. Consider the Dirichlet problem on $\Delta := \Delta(w, \rho)$ with $\phi = h|_{\partial \Delta}$. By Theorem 1.2.4, h and $P_\Delta h$ are both solutions, so by Theorem 1.2.2, $h = P_\Delta h$ on Δ. □

Note that this result is a generalization of the mean-value property, which is just the case $r = 0$. It allows us to recapture the values of h everywhere on Δ from knowledge of h on $\partial \Delta$. Exercise 4 gives an analogous formula for f on Δ, where f is the essentially unique holomorphic function such that $h = \operatorname{Re} f$.

The mean-value property actually characterizes harmonic functions. This is proved in the next theorem, which also illustrates well the value of being able to solve the Dirichlet problem.

Theorem 1.2.7 (Converse to Mean-Value Property) *Let $h: U \to \mathbb{R}$ be a continuous function on an open subset U of $\mathbb{C}$, and suppose that it possesses the local mean-value property, i.e. given $w \in U$, there exists $\rho > 0$ such that*

$$h(w) = \frac{1}{2\pi} \int_0^{2\pi} h(w + re^{it}) \, dt \quad (0 \leq r < \rho).$$

Then h is harmonic on D.

1.2. THE DIRICHLET PROBLEM ON THE DISC

Proof. It is enough to show that h is harmonic on each open disc Δ with $\overline{\Delta} \subset U$. Fix such a Δ, and define $k \colon \overline{\Delta} \to \mathbb{R}$ by

$$k = \begin{cases} h - P_\Delta h, & \text{on } \Delta, \\ 0, & \text{on } \partial\Delta. \end{cases}$$

Then k is continuous on $\overline{\Delta}$ and has the local mean-value property on Δ. As $\overline{\Delta}$ is compact, k attains a maximum value M at some point of $\overline{\Delta}$. Define

$$A = \{z \in \Delta : k(z) < M\} \quad \text{and} \quad B = \{z \in \Delta : k(z) = M\}.$$

Then A is open, since k is continuous. Also B is open, for if $k(w) = M$, then the local mean-value property forces k to be equal to M on all sufficiently small circles around w. As A and B partition the connected set Δ, either $A = \Delta$, in which case k attains its maximum on $\partial\Delta$ and so $M = 0$, or else $B = \Delta$, in which case $k \equiv M$ and again $M = 0$. Thus $k \leq 0$, and a similar argument shows that $k \geq 0$. Hence $h = P_\Delta h$ on Δ, and since $P_\Delta h$ is harmonic there, so is h. □

Corollary 1.2.8 *If $(h_n)_{n \geq 1}$ is a sequence of harmonic functions on D converging locally uniformly to a function h, then h is also harmonic on D.*

Proof. Combine Theorems 1.1.6 and 1.2.7. □

A useful feature of Theorem 1.2.7 is that one only needs to check that the mean-value property holds *locally* (i.e. the value of ρ can depend upon w). As an application of this, we derive a form of the reflection principle for holomorphic functions.

Theorem 1.2.9 (Reflection Principle) *Let $\Delta = \Delta(0, R)$, and write*

$$\Delta^+ = \{z \in \Delta : \operatorname{Im} z > 0\}, \quad \text{and} \quad I = \{z \in \Delta : \operatorname{Im} z = 0\}.$$

Suppose that f is a holomorphic function on Δ^+ such that $\operatorname{Re} f$ extends continuously to $\Delta^+ \cup I$ with $\operatorname{Re} f = 0$ on I. Then f extends holomorphically to the whole of Δ.

Note that no assumption is made about continuity of $\operatorname{Im} f$ on I—this comes for free.

Proof. Define $h \colon \Delta \to \mathbb{R}$ by

$$h(z) = \begin{cases} \operatorname{Re} f(z), & z \in \Delta^+, \\ 0, & z \in I, \\ -\operatorname{Re} f(\overline{z}), & \overline{z} \in \Delta^+. \end{cases}$$

Then h is continuous on Δ and has the local mean-value property there, so by Theorem 1.2.7 it is harmonic on Δ. From Theorem 1.1.2(b), there exists a holomorphic function $\tilde{f}$ on Δ such that $h = \operatorname{Re} \tilde{f}$. Now $f - \tilde{f}$ is holomorphic on Δ^+ and takes only imaginary values, so it is constant there. Adjusting $\tilde{f}$ appropriately, we can make this constant zero. Then $\tilde{f}$ provides the promised holomorphic extension of f to the whole of Δ. □

Exercises 1.2

1. Let $\Delta = \Delta(0,1)$ and define $\phi\colon \partial\Delta \to \mathbb{C}$ by $\phi(\zeta) = \overline{\zeta}$. Show that there is no function f holomorphic on Δ such that $\lim_{z\to\zeta} f(z) = \phi(\zeta)$ for all $\zeta \in \partial\Delta$. [Thus the holomorphic version of the Dirichlet problem on the disc may have no solution.]

2. (i) Show that the Poisson kernel is given by
$$P(re^{it}, e^{i\theta}) = \sum_{n\in\mathbb{Z}} r^{|n|} e^{in(t-\theta)} \quad (r<1,\ 0 \le t, \theta < 2\pi).$$
Use this to give an alternative proof of Lemma 1.2.5 (ii).

   (ii) Show that if $\phi\colon \partial\Delta(0,1) \to \mathbb{R}$ is an integrable function, then
$$P_\Delta \phi(re^{it}) = \sum_{n\in\mathbb{Z}} a_n r^{|n|} e^{int} \quad (r<1,\ 0 \le t < 2\pi),$$
where $(a_n)_{n\in\mathbb{Z}}$ is a bounded sequence of real numbers.

   (iii) Assume now that ϕ is continuous. Writing $\phi_r(e^{it}) = P_\Delta \phi(re^{it})$, show that $\phi_r \to \phi$ uniformly on $\partial\Delta(0,1)$ as $r \to 1$, and deduce that $\phi(e^{it})$ can be uniformly approximated by trigonometric polynomials $\sum_{-N}^{N} b_n e^{int}$.

3. Let $(h_n)_{n\ge 1}$ be a sequence of harmonic functions on a domain D converging locally uniformly to a (harmonic) function h. Show that $(h_n)_x$ and $(h_n)_y$ converge locally uniformly to h_x and h_y respectively.

4. Show that if f is holomorphic on a neighbourhood of $\overline{\Delta}(0,\rho)$, and $h = \operatorname{Re} f$, then
$$f(z) - f(0) = \frac{1}{2\pi} \int_0^{2\pi} \frac{2z}{\rho e^{i\theta} - z} h(\rho e^{i\theta})\, d\theta \quad (|z| < \rho).$$
[Hint: first show that the real parts agree.]

5. Show that the holomorphic extension $\tilde{f}$ of f in the reflection principle (Theorem 1.2.9) is given by
$$\tilde{f}(\overline{z}) = -\overline{f(z)} \quad (\overline{z} \in \Delta^+).$$

1.3 Positive Harmonic Functions

In this section we shall exploit the Poisson integral formula to derive some useful inequalities for positive harmonic functions. By 'positive' here is meant 'non-negative', although in this context there is hardly any difference since, by Theorem 1.1.8, any harmonic function which attains a minimum value zero on a domain must be identically zero throughout the domain.

Theorem 1.3.1 (Harnack's Inequality) *Let h be a positive harmonic function on the disc $\Delta(w, \rho)$. Then for $r < \rho$ and $0 \le t < 2\pi$,*

$$\frac{\rho - r}{\rho + r} h(w) \le h(w + re^{it}) \le \frac{\rho + r}{\rho - r} h(w).$$

Proof. Choose s with $r < s < \rho$. By the Poisson integral formula applied to h on $\overline{\Delta}(w, s)$,

$$\begin{aligned}
h(w + re^{it}) &= \frac{1}{2\pi} \int_0^{2\pi} \frac{s^2 - r^2}{s^2 - 2sr\cos(\theta - t) + r^2} h(w + se^{i\theta}) \, d\theta \\
&\le \frac{1}{2\pi} \int_0^{2\pi} \frac{s + r}{s - r} h(w + se^{i\theta}) \, d\theta \\
&= \frac{s + r}{s - r} h(w),
\end{aligned}$$

the last equality being just the mean-value property for h. Letting $s \to \rho$, we deduce that

$$h(w + re^{it}) \le \frac{\rho + r}{\rho - r} h(w),$$

which is the desired upper bound. The lower bound is proved in a similar fashion. □

Corollary 1.3.2 (Liouville Theorem) *Every harmonic function on $\mathbb{C}$ which is bounded above or below is constant.*

Proof. It is enough to show that every positive harmonic function h on $\mathbb{C}$ is constant. Given $z \in \mathbb{C}$, put $r = |z|$ and let $\rho > r$. Applying Harnack's inequality to h on $\Delta(0, \rho)$ gives

$$h(z) \le \frac{\rho + r}{\rho - r} h(0).$$

Letting $\rho \to \infty$, we deduce that $h(z) \le h(0)$. Thus h attains a maximum at 0, and by Theorem 1.1.8 this implies that h is constant. □

Harnack's inequality on discs implies an analogous result for general domains.

Corollary 1.3.3 *Let D be a domain in $\mathbb{C}_\infty$ and let $z, w \in D$. Then there exists a number τ such that, for every positive harmonic function h on D,*

$$\tau^{-1}h(w) \leq h(z) \leq \tau h(w). \tag{1.2}$$

Proof. Given $z, w \in D$, write $z \sim w$ if there exists a number τ such that (1.2) holds for every positive harmonic function h on D. Then $\sim$ is an equivalence relation on D, and Harnack's inequality shows that the equivalence classes are open sets. As D is connected, there can only be one such equivalence class, and this proves the corollary. $\square$

Prompted by this last result, we make the following definition.

Definition 1.3.4 *Let D be a domain in $\mathbb{C}_\infty$. Given $z, w \in D$, the Harnack distance between z and w is the smallest number $\tau_D(z, w)$ such that, for every positive harmonic function h on D,*

$$\tau_D(z,w)^{-1} h(w) \leq h(z) \leq \tau_D(z,w) h(w). \tag{1.3}$$

There is one case for which τ_D can be computed straightaway.

Theorem 1.3.5 *If $\Delta = \Delta(w, \rho)$, then*

$$\tau_\Delta(z, w) = \frac{\rho + |z - w|}{\rho - |z - w|} \quad (z \in \Delta).$$

Proof. From Harnack's inequality, it follows that

$$\tau_\Delta(z, w) \leq \frac{\rho + |z - w|}{\rho - |z - w|} \quad (z \in \Delta).$$

On the other hand, by considering the positive harmonic functions h on Δ given by

$$h(z) = P\left(\frac{z-w}{\rho}, \zeta\right) = \operatorname{Re}\left(\frac{\rho\zeta + (z-w)}{\rho\zeta - (z-w)}\right) \quad (|\zeta| = 1),$$

we see that in fact equality holds. $\square$

From this, one can compute or estimate τ_D for other domains D by means of the following subordination principle.

1.3. POSITIVE HARMONIC FUNCTIONS

Theorem 1.3.6 (Subordination Principle) *Let $f\colon D_1 \to D_2$ be a meromorphic map between domains D_1 and D_2 in $\mathbf{C}_\infty$. Then*

$$\tau_{D_2}(f(z), f(w)) \leq \tau_{D_1}(z, w) \quad (z, w \in D_1),$$

with equality if f is a conformal mapping of D_1 onto D_2.

Proof. Let $z, w \in D_1$. Given a positive harmonic function h on D_2, the composite $h \circ f$ is a positive harmonic function on D_1, so from (1.3)

$$\tau_{D_1}(z, w)^{-1} h(f(w)) \leq h(f(z)) \leq \tau_{D_1}(z, w) h(f(w)).$$

As this holds for every such h, the desired inequality follows.

If f is a conformal map of D_1 onto D_2, then we can apply the same argument to f^{-1} to deduce that equality holds. □

Corollary 1.3.7 *If $D_1 \subset D_2$ then $\tau_{D_2}(z, w) \leq \tau_{D_1}(z, w)$ $(z, w \in D_1)$.*

Proof. Take $f\colon D_1 \to D_2$ to be the inclusion map. □

We can use this to study the continuity properties of τ_D.

Theorem 1.3.8 *If D is a subdomain of $\mathbf{C}_\infty$, then $\log \tau_D$ is a continuous semimetric on D.*

Proof. To show that $\log \tau_D$ is a semimetric, we need to check that

$$\begin{array}{ll} \tau_D(z, w) \geq 1, \quad \tau_D(z, z) = 1 & (z, w \in D), \\ \tau_D(z, w) = \tau_D(w, z) & (z, w \in D), \\ \tau_D(z, w) \leq \tau_D(z, z') \tau_D(z', w) & (z, z', w \in D), \end{array}$$

all of which follow easily from the definition of τ_D.

To show that $\log \tau_D$ is continuous, it is sufficient to prove that

$$\log \tau_D(z, w) \to 0 \quad \text{as } z \to w,$$

because the general result then follows by the triangle inequality for $\log \tau_D$. To this end, let $w \in D$, and choose $\rho > 0$ so that $\Delta := \Delta(w, \rho) \subset D$. Then for $z \in \Delta$ we have

$$0 \leq \log \tau_D(z, w) \leq \log \tau_\Delta(z, w) = \log\left(\frac{\rho + |z - w|}{\rho - |z - w|}\right),$$

whence indeed $\log \tau_D(z, w) \to 0$ as $z \to w$. □

It may happen that $\log \tau_D(z,w) = 0$ even when $z \neq w$, so that $\log \tau_D$ is not quite a metric. For example, since the only positive harmonic functions on $\mathbb{C}$ are constants, it follows that $\log \tau_\mathbb{C}(z,w) = 0$ for all $z, w \in \mathbb{C}$. However, $\log \tau_D$ *is* a metric for many domains D: this is pursued a little further in Exercise 4.

It is now only a short step to the following important theorem.

Theorem 1.3.9 (Harnack's Theorem) *Let $(h_n)_{n \geq 1}$ be harmonic functions on a domain D in $\mathbb{C}_\infty$, and suppose that $h_1 \leq h_2 \leq h_3 \leq \cdots$ on D. Then either $h_n \to \infty$ locally uniformly, or else $h_n \to h$ locally uniformly, where h is harmonic on D.*

Proof. Fix $w \in D$. Given a compact subset K of D, the quantity

$$C_K := \sup_{z \in K} \tau_D(z, w)$$

is finite, since τ_D is continuous. Hence whenever $n \geq m \geq 1$, we have

$$h_n(w) - h_1(w) \leq C_K(h_n(z) - h_1(z)) \quad (z \in K),$$
$$h_n(z) - h_m(z) \leq C_K(h_n(w) - h_m(w)) \quad (z \in K),$$

because $h_n - h_1$ and $h_n - h_m$ are positive harmonic functions on D. Now if $h_n(w) \to \infty$ as $n \to \infty$, it follows that $h_n \to \infty$ uniformly on K. As K can be any compact subset of D, we conclude that $h_n \to \infty$ locally uniformly on D. On the other hand, if $h_n(w)$ tends to a finite limit, then $(h_n)_{n \geq 1}$ is uniformly Cauchy on K. Again, as K is arbitrary, it follows that h_n converges locally uniformly on D to a finite function h, which must be harmonic on D by Corollary 1.2.8. □

There is also a useful variant of Harnack's theorem in which, instead of assuming that the sequence (h_n) is increasing, we suppose merely that it is positive. The price we pay is that, in general, only a subsequence will converge.

Theorem 1.3.10 *Let $(h_n)_{n \geq 1}$ be positive harmonic functions on a domain D in $\mathbb{C}_\infty$. Then either $h_n \to \infty$ locally uniformly, or else some subsequence $h_{n_j} \to h$ locally uniformly, where h is harmonic on D.*

Proof. Fix $w \in D$. From the inequalities

(1.4) $\quad \tau_D(z,w)^{-1} h_n(w) \leq h_n(z) \leq \tau_D(z,w) h_n(w) \quad (z \in D, \ n \geq 1),$

it follows that if $h_n(w) \to \infty$, then also $h_n \to \infty$ locally uniformly on D, and if $h_n(w) \to 0$, then also $h_n \to 0$ locally uniformly on D. Therefore,

1.3. POSITIVE HARMONIC FUNCTIONS

replacing (h_n) by a subsequence if necessary, we can reduce to the case where the sequence $(\log h_n(w))_{n\geq 1}$ is bounded. The inequality (1.4) then implies that $(\log h_n)_{n\geq 1}$ is locally uniformly bounded on D, and so it suffices to prove that there is a subsequence (h_{n_j}) such that $(\log h_{n_j})_{j\geq 1}$ is locally uniformly convergent on D.

Let S be a countable dense subset of D. The sequence $(\log h_n(\zeta))_{n\geq 1}$ is bounded for each $\zeta \in S$, so by a 'diagonal argument' we may find a subsequence (h_{n_j}) such that $(\log h_{n_j}(\zeta))_{j\geq 1}$ is convergent for each $\zeta \in S$. We shall show that, for this subsequence, $(\log h_{n_j})_{j\geq 1}$ is locally uniformly convergent on D.

Let K be a compact subset of D, and let $\epsilon > 0$. For each $z \in K$, let

$$V_z = \{z' \in D : \log \tau_D(z,z') < \epsilon\},$$

and let $V_{z_1}, \ldots, V_{z_m}$ be a finite subcover of K. As S is dense in D, for each l we can pick a point $\zeta_l \in V_{z_l} \cap S$. Then there exists $N \geq 1$ such that

$$|\log h_{n_j}(\zeta_l) - \log h_{n_k}(\zeta_l)| \leq \epsilon \quad (n_j, n_k \geq N, \ l = 1, \ldots, m).$$

By definition of the Harnack distance,

$$|\log h_{n_j}(z) - \log h_{n_j}(\zeta_l)| \leq \log \tau_D(z, \zeta_l) < 2\epsilon \quad (z \in V_{z_l}),$$

with a similar inequality for h_{n_k}. Hence

$$|\log h_{n_j}(z) - \log h_{n_k}(z)| < 5\epsilon \quad (n_j, n_k \geq N, \ z \in K).$$

Thus $(\log h_{n_j})_{j\geq 1}$ is uniformly Cauchy on K, and so uniformly convergent there. $\square$

We conclude this chapter by applying some of the ideas developed in it to give a beautiful recent proof of Picard's theorem due to John Lewis.

Theorem 1.3.11 (Picard's Theorem) *If $f: \mathbb{C} \to \mathbb{C}$ is a non-constant entire function, then $\mathbb{C} \setminus f(\mathbb{C})$ contains at most one point.*

The proof requires a lemma on harmonic functions which is of some interest in its own right. We shall use the notation

$$M_h(w,r) := \sup_{\Delta(w,r)} h = \sup_{\partial \Delta(w,r)} h.$$

Lemma 1.3.12 *Let h be harmonic on a neighbourhood of $\overline{\Delta}(0, 2R)$ with $h(0) = 0$. Then there exists a disc $\Delta(w,r) \subset \Delta(0, 2R)$ such that $h(w) = 0$ and*

$$\begin{aligned} M_h(w,r) &\geq 3^{-11} M_h(0,R), \\ M_h(w,r/2) &\geq 3^{-11} M_h(w,r). \end{aligned}$$

Of course the exact value of the constant 3^{-11} is unimportant here. The point is that it is positive!

Proof. For $z \in \Delta(0, 2R)$ write $\delta(z) = \text{dist}(z, \partial\Delta(0, 2R))$, and define

$$\begin{aligned} Z &= \{z \in \Delta(0, 2R) : h(z) = 0\}, \\ U &= \bigcup_{z \in Z} \Delta(z, \delta(z)/4), \\ \gamma &= \sup_U h = \sup_{z \in Z} M_h(z, \delta(z)/4). \end{aligned}$$

Choose $w \in Z$ such that $M_h(w, \delta(w)/4) \geq \gamma/3$, and set $r = \delta(w)/2$. We shall show that $\Delta(w, r)$ satisfies the conclusions of the lemma. Clearly $\Delta(w, r) \subset \Delta(0, 2R)$ and $h(w) = 0$. Also $M_h(w, r/2) \geq \gamma/3$, so to complete the proof it suffices to show that

(a) $M_h(0, R) \leq 3^{10}\gamma$,
(b) $M_h(w, r) \leq 3^{10}\gamma$.

Proof of (a): Take $z \in \Delta(0, R)$ with $h(z) \geq 0$. If $z \in \overline{U}$, then by continuity $h(z) \leq \gamma$. Now suppose that $z \notin \overline{U}$. Then (using the obvious notation for line segments in $\mathbb{C}$) there exists $z' \in (z, 0) \cap \overline{U}$ such that $[z, z') \cap \overline{U} = \emptyset$. It follows that $h > 0$ on $[z, z')$. In fact, for each $\zeta \in [z, z')$ we have $h > 0$ on $\Delta(\zeta, R/5)$. (For if not, then there exists $\zeta' \in \Delta(\zeta, R/5)$ with $h(\zeta') = 0$. But then $\zeta' \in Z$ and

$$\delta(\zeta') \geq \delta(\zeta) - |\zeta' - \zeta| \geq R - R/5 = 4R/5 > 4|\zeta' - \zeta|,$$

implying that $\zeta \in U$, a contradiction.) It follows by Harnack's inequality that, for each such ζ,

$$\sup_{\Delta(\zeta, R/10)} h \leq 3^2 \inf_{\Delta(\zeta, R/10)} h.$$

Since $[z, z']$ has length less than R, it can be covered by 5 overlapping discs of radius $R/10$ with centres in $[z, z']$. Therefore

$$h(z) \leq (3^2)^5 h(z') \leq 3^{10}\gamma,$$

the final inequality holding because $h \leq \gamma$ on $\overline{U}$.

Proof of (b): This is virtually identical. Take $z \in \Delta(w, r)$ with $h(z) \geq 0$. If $z \in \overline{U}$, then by continuity $h(z) \leq \gamma$. Now suppose that $z \notin \overline{U}$. Then there exists $z' \in (z, w) \cap \overline{U}$ such that $[z, z') \cap \overline{U} = \emptyset$. It follows that $h > 0$

1.3. POSITIVE HARMONIC FUNCTIONS

on $[z, z']$. In fact, for each $\zeta \in [z, z']$ we have $h > 0$ on $\Delta(\zeta, r/5)$. (For if not, then there exists $\zeta' \in \Delta(\zeta, r/5)$ with $h(\zeta') = 0$. But then $\zeta' \in Z$ and

$$\delta(\zeta') \geq \delta(w) - |\zeta' - \zeta| - |\zeta - w| \geq 2r - r/5 - r = 4r/5 > 4|\zeta' - \zeta|,$$

implying that $\zeta \in U$, a contradiction.) It follows by Harnack's inequality that, for each such ζ,

$$\sup_{\Delta(\zeta, r/10)} h \leq 3^2 \inf_{\Delta(\zeta, r/10)} h.$$

Since $[z, z']$ has length less than r, it can be covered by 5 overlapping discs of radius $r/10$ with centres in $[z, z')$. Therefore

$$h(z) \leq (3^2)^5 h(z') \leq 3^{10} \gamma,$$

the final inequality holding, as before, because $h \leq \gamma$ on $\overline{U}$. $\square$

Proof of Theorem 1.3.11. Suppose, for a contradiction, that $\mathbb{C} \setminus f(\mathbb{C})$ contains at least two points, α, β. Then $h := \log|f - \alpha|$ and $k := \log|f - \beta|$ are both harmonic functions on $\mathbb{C}$, and they satisfy

$$\begin{aligned}|h^+ - k^+| &\leq |\alpha - \beta|, \\ \max(h, k) &\geq \log(|\alpha - \beta|/2),\end{aligned}$$

everywhere on $\mathbb{C}$. Since f is non-constant, so is h, and so by Theorem 1.3.2 h is unbounded above and below. In particular, there exists $z_0 \in \mathbb{C}$ with $h(z_0) = 0$, and replacing $f(z)$ by $f(z + z_0)$ we can suppose without loss of generality that $z_0 = 0$. We now apply Lemma 1.3.12 to h on each of the discs $\Delta(0, 2^{j+1})$ to produce new discs $\Delta(w_j, r_j)$ such that $h(w_j) = 0$ and

$$\begin{aligned}M_h(w_j, r_j) &\geq 3^{-11} M_h(0, 2^j), \\ M_h(w_j, r_j/2) &\geq 3^{-11} M_h(w_j, r_j).\end{aligned}$$

For each $j \geq 1$ set $M_j = M_h(w_j, r_j)$. Since h is unbounded,

$$\lim_{j \to \infty} M_j \geq 3^{-11} \lim_{j \to \infty} M_h(0, 2^j) = \infty.$$

Define two sequences of harmonic functions (h_j) and (k_j) on $\Delta(0, 1)$ by

$$h_j(z) = \frac{h(w_j + r_j z)}{M_j} \quad \text{and} \quad k_j(z) = \frac{k(w_j + r_j z)}{M_j} \quad (|z| < 1).$$

Then h_j and k_j have the following properties:

(a) $h_j(0) = 0$,
(b) $M_{h_j}(0, \frac{1}{2}) \geq 3^{-11}$,
(c) $|h_j^+ - k_j^+| \leq |\alpha - \beta|/M_j$,
(d) $\max(h_j, k_j) \geq \log(|\alpha - \beta|/2)/M_j$.

Evidently $h_j \leq 1$ for all j, so we can apply Theorem 1.3.10 to $(1 - h_j)_{j \geq 1}$ to deduce that a subsequence of the (h_j) converges locally uniformly to a function $\widetilde{h}$ on $\Delta(0, 1)$. The (k_j) are also uniformly bounded above (for example by $1 + |\alpha - \beta|/M_1$), and so a further subsequence of these converges locally uniformly to a function $\widetilde{k}$ on $\Delta(0, 1)$. Both $\widetilde{h}$ and $\widetilde{k}$ are harmonic (or possibly identically $-\infty$), and they have the following properties:

(a) $\widetilde{h}(0) = 0$,
(b) $M_{\widetilde{h}}(0, \frac{1}{2}) \geq 3^{-11}$,
(c) $\widetilde{h}^+ = \widetilde{k}^+$
(d) $\max(\widetilde{h}, \widetilde{k}) \geq 0$.

Property (b) implies that $\widetilde{h}(\zeta) > 0$ for some ζ, and (c) then tells us that $\widetilde{h} = \widetilde{k}$ in a neighbourhood of ζ. By the identity principle (Theorem 1.1.7) it follows that $\widetilde{h} = \widetilde{k}$ everywhere on $\Delta(0, 1)$. From (d) we then deduce that $\widetilde{h} \geq 0$ on $\Delta(0, 1)$, and combining this with (a) and the maximum principle (Theorem 1.1.8), we conclude that $\widetilde{h} \equiv 0$ on $\Delta(0, 1)$. But this is inconsistent with (b)! We have thus arrived at our contradiction, and Picard's theorem is proved. □

Exercises 1.3

1. Show that if h is a positive harmonic function on $\Delta(0, \rho)$ then

$$|\nabla h(0)| \leq \frac{2}{\rho} h(0),$$

and hence deduce that

$$|\nabla h(z)| \leq \frac{2\rho}{\rho^2 - |z|^2} h(z) \quad (|z| < \rho).$$

2. Show that if f is a holomorphic function on $\Delta(0, \rho)$ which satisfies $0 < |f| < 1$, then

$$|f(z)| \leq |f(0)|^{(\rho - |z|)/(\rho + |z|)} \quad (|z| < \rho).$$

1.3. POSITIVE HARMONIC FUNCTIONS

3. Show that if $\Delta = \Delta(0,1)$ then
$$\tau_\Delta(z,w) = \frac{1 + \left|\frac{z-w}{1-z\bar{w}}\right|}{1 - \left|\frac{z-w}{1-z\bar{w}}\right|} \qquad (z, w \in \Delta).$$

4. Let D be a bounded domain in $\mathbb{C}$ of diameter δ.

   (i) Show that
   $$\tau_D(z,w) \geq \frac{\delta + |z-w|}{\delta - |z-w|} \qquad (z, w \in D),$$
   and deduce that $\log \tau_D$ is a metric on D giving rise to the usual topology on D.

   (ii) Show that if $w \in D$ and $\zeta \in \partial D$ then
   $$\tau_D(z,w) \to \infty \quad \text{as } z \to \zeta,$$
   and deduce that the metric space $(D, \log \tau_D)$ is complete.

   Show also that the final conclusions of (i) and (ii) remain valid if D is any domain in $\mathbb{C}_\infty$ conformally equivalent to a bounded domain.

5. Let h be a function harmonic on a neighbourhood of $\overline{\Delta}(0, \rho)$, and for $0 \leq r \leq \rho$ define
$$M_h(r) = \sup_{|z|=r} h(z).$$
Prove that
$$M_h(r) \leq \frac{2r}{\rho+r} M_h(\rho) + \frac{\rho-r}{\rho+r} h(0) \qquad (0 \leq r \leq \rho).$$
Deduce the following generalization of the Liouville theorem (Corollary 1.3.2): if h is harmonic on $\mathbb{C}$ and satisfies
$$\liminf_{\rho \to \infty} \frac{M_h(\rho)}{\rho} \leq 0,$$
then h is constant.

6. Let f be a function holomorphic on a neighbourhood of $\overline{\Delta}(0, \rho)$, and for $0 \leq r \leq \rho$ define
$$M_f(r) = \sup_{|z|=r} |f(z)| \quad \text{and} \quad A_f(r) = \sup_{|z|=r} \operatorname{Re} f(z).$$
Prove the Borel–Carathéodory inequality:
$$M_f(r) \leq \frac{2r}{\rho-r} A_f(\rho) + \frac{\rho+r}{\rho-r} |f(0)| \qquad (0 \leq r < \rho).$$
[Hint: apply the result of Exercise 1.2.4 to the function $(A_f(\rho) - f)$.]

7. Let $(h_n)_{n \geq 1}$ be positive harmonic functions on a domain D. Show that if $(h_n(z))_{n \geq 1}$ converges pointwise for each z in some non-empty open subset of D, then $(h_n)_{n \geq 1}$ converges locally uniformly on the whole of D.

8. (i) Let h and k be harmonic functions on a neighbourhood of $\overline{\Delta}(0,1)$ satisfying

$$|h(0)|, |k(0)| \leq \lambda,$$
$$|h^+ - k^+| \leq \lambda,$$
$$\max(h, k) \geq -\lambda,$$

for some constant λ. Use the technique in Theorem 1.3.11 to show that there is a universal constant c_0 such that

$$\sup_{\Delta(0,1/2)} h \leq c_0 \lambda \quad \text{and} \quad \sup_{\Delta(0,1/2)} k \leq c_0 \lambda.$$

(ii) Deduce Montel's theorem: if (f_n) is a sequence of holomorphic functions on a domain D such that $f_n(D) \subset \mathbb{C} \setminus \{0, 1\}$ for all n, then either $f_n \to \infty$ locally uniformly, or else some subsequence $f_{n_j} \to f$ locally uniformly, where f is holomorphic on D.

Notes on Chapter 1

§1.1

The same definition of harmonic function, as a solution of Laplace's equation, makes equally good sense in $\mathbb{R}^n$ for all n. Much of theory developed in this book carries over to this case, though there are also important differences—see for example the books of Hayman and Kennedy [34] and Helms [36]. However, for applications to complex variable, other generalizations may be more appropriate. For instance, the definition of harmonicity on the Riemann sphere, outlined after Corollary 1.1.5, also works on an arbitrary Riemann surface, and in fact harmonic functions play a key rôle in the analysis of Riemann surfaces. An introductory account of this theory can be found in Beardon [12]. Also the analogue of Corollary 1.1.5 breaks down for harmonic functions in dimensions higher than 2, so in studying several complex variables it is more natural to consider the so-called pluriharmonic functions, a subclass which is invariant under holomorphic maps. The corresponding theory, pluripotential theory, is described in the recent book of Klimek [38].

§1.3

The proof given of Picard's theorem (Theorem 1.3.11) is due to Lewis [42], who simplified an earlier proof of Eremenko and Lewis [27]. Exercise 8 is also drawn from these sources. For entire functions of finite class, there is a yet more elementary proof of Picard's theorem, due to Borel, based on the Borel–Carathéodory inequality (Exercise 6). For further details of this see for example [14, Proposition 4.5.12].

Chapter 2

Subharmonic Functions

2.1 Upper Semicontinuous Functions

As part of their definition, subharmonic functions are going to be upper semicontinuous, so before making this definition, we take a brief look at upper semicontinuous functions in the abstract.

Definition 2.1.1 Let X be a topological space. We say that a function $u\colon X \to [-\infty, \infty)$ is *upper semicontinuous* if the set $\{x \in X : u(x) < \alpha\}$ is open in X for each $\alpha \in \mathbb{R}$. Also $v\colon X \to (-\infty, \infty]$ is *lower semicontinuous* if $-v$ is upper semicontinuous.

A straightforward check shows that u is upper semicontinuous if and only if
$$\limsup_{y \to x} u(y) \leq u(x) \quad (x \in X).$$
In partic Also u is continuous if and only if it is both upper and lower semicontinuous.

We shall make frequent use of the following basic compactness theorem.

Theorem 2.1.2 *Let u be an upper semicontinuous function on a topological space X, and let K be a compact subset of X. Then u is bounded above on K and attains its bound.*

Proof. The sets $\{x \in X : u(x) < n\}$ $(n \geq 1)$ form an open cover of K, so have a finite subcover. Hence u is bounded above on K. Let $M = \sup_K u$. Then the open sets $\{x \in X : u(x) < M - 1/n\}$ $(n \geq 1)$ cannot cover K, because they have no finite subcover. Hence $u(x) = M$ for at least one $x \in K$. □

The other result that we shall need is an approximation theorem.

Theorem 2.1.3 *Let u be an upper semicontinuous function on a metric space (X,d), and suppose that u is bounded above on X. Then there exist continuous functions $\phi_n : X \to \mathbb{R}$ such that $\phi_1 \geq \phi_2 \geq \cdots \geq u$ on X and $\lim_{n\to\infty} \phi_n = u$.*

Proof. We can suppose that $u \not\equiv -\infty$ (otherwise just take $\phi_n \equiv -n$). For $n \geq 1$, define $\phi_n : X \to \mathbb{R}$ by
$$\phi_n(x) = \sup_{y \in X}(u(y) - nd(x,y)) \quad (x \in X).$$
Then for each n we have
$$|\phi_n(x) - \phi_n(x')| \leq nd(x,x') \quad (x, x' \in X),$$
so ϕ_n is continuous on X. Clearly also $\phi_1 \geq \phi_2 \geq \cdots \geq u$, and so in particular $\lim_{n\to\infty} \phi_n \geq u$. On the other hand, writing $\Delta(x,\rho)$ for the ball $\{y \in X : d(x,y) < \rho\}$, we have
$$\phi_n(x) \leq \max\left(\sup_{\Delta(x,\rho)} u, \sup_X u - n\rho\right) \quad (x \in X, \rho > 0),$$
so that
$$\lim_{n\to\infty} \phi_n(x) \leq \sup_{\Delta(x,\rho)} u \quad (x \in X, \rho > 0).$$
As u is upper semicontinuous, letting $\rho \to 0$ gives $\lim_{n\to\infty} \phi_n \leq u$. □

Exercises 2.1

1. Let S be a subset of a topological space X. Show that 1_S, the characteristic function of S, is upper semicontinuous if and only if S is closed in X.

2. Let u be an upper semicontinuous function on a metric space (X,d). For $n \geq 0$ define
$$\begin{aligned} F_n &= \{x \in X : u(x) \geq n\}, \\ \psi_n(x) &= \max(0, 1 - n\,\mathrm{dist}(x, F_n)) \quad (x \in X). \end{aligned}$$
Show that $\sum_{n \geq 0} \psi_n$ converges locally uniformly to a continuous function $\psi : X \to \mathbb{R}$ such that $\psi \geq u$ on X. By considering $u - \psi$, deduce that Theorem 2.1.3 remains valid without the assumption that u is bounded above on X.

2.2. SUBHARMONIC FUNCTIONS

3. Let (X_1, d_1) and (X_2, d_2) be metric spaces, and let $f: X_1 \to X_2$ be an arbitrary function. Given $x \in X_1$, the *oscillation* of f at x is defined by

$$\omega_f(x) = \lim_{\rho \to 0} \Big(\sup\{d_2(f(y), f(y')) : d_1(x, y) < \rho, d_1(x, y') < \rho\} \Big).$$

   (i) Show that f is continuous at x if and only if $\omega_f(x) = 0$.

   (ii) Show that the set $\{x : \omega_f(x) < \alpha\}$ is open in X_1 for each $\alpha > 0$, and deduce that the points at which f is continuous form a G_δ subset of X_1.

   (iii) Suppose also that (X_1, d_1) is complete, and that f is the pointwise limit of a sequence of continuous functions $f_n: X_1 \to X_2$. Show that the set $\{x : \omega_f(x) < \alpha\}$ is dense in X_1 for each $\alpha > 0$, and deduce that the points at which f is continuous form a dense G_δ subset of X_1. [Hint: use the Baire category theorem.]

Conclude that a function upper semicontinuous on a complete metric space is continuous at a dense G_δ set of points.

2.2 Subharmonic Functions

In spirit, at least, a function u is subharmonic if its Laplacian satisfies $\Delta u \geq 0$. However, we shall not define subharmonicity this way. As we shall see later, one of the great virtues of subharmonic functions is their flexibility, and this would be lost if we were to assume that they were smooth.

Instead, we proceed by analogy with convex functions on $\mathbb{R}$ (indeed, this is a good analogy to keep in mind throughout the book). If $\psi \in C^2(\mathbb{R})$, then it is convex if and only if $\psi'' \geq 0$, but convexity is actually *defined* via a submean property (see Section 2.6), which also allows non-smooth functions such as $\psi(t) = |t|$ to be convex. Taking this as our model, we shall define subharmonicity using an analogous submean property in the plane.

There is, however, one more technicality. Convex functions on open intervals are automatically continuous, but there is no such result for subharmonic functions. We could demand continuity as part of the definition, but, for reasons that will become apparent later, it is advantageous merely to ask for upper semicontinuity.

After this preamble, we are at last ready to make the definition.

Definition 2.2.1 Let U be an open subset of $\mathbb{C}$. A function $u\colon U \to [-\infty, \infty)$ is called *subharmonic* if it is upper semicontinuous and satisfies the *local submean inequality*, i.e. given $w \in U$, there exists $\rho > 0$ such that

$$(2.1) \qquad u(w) \leq \frac{1}{2\pi} \int_0^{2\pi} u(w + re^{it})\,dt \qquad (0 \leq r < \rho).$$

Also $v\colon U \to (-\infty, \infty]$ is *superharmonic* if $-v$ is subharmonic.

This definition merits some comment. First of all, the integral in (2.1) is to be interpreted as the difference of the corresponding integrals of u^+ and u^-. By Theorem 2.1.2, u^+ is bounded on $\partial \Delta(w, r)$, so its integral is certainly finite. Thus the difference of the two integrals makes sense, even though the integral of u^- may be infinite. We shall see later that in fact the latter only happens when $u \equiv -\infty$ on the whole component of U containing w. (Note that, according to our definition, $u \equiv -\infty$ *is* a subharmonic function, though some authors do exclude it.)

The second remark is that because subharmonicity is defined via the *local* submean inequality (i.e. ρ may depend on w), it is a local property. This means that if (U_α) is an open cover of U, then u is subharmonic on U if and only if it is subharmonic on each U_α.

Thirdly, we observe that a function is harmonic if and only if it is both subharmonic and superharmonic. One way this follows from Theorem 1.1.6, and the other way from Theorem 1.2.7.

We now give some examples of subharmonic functions.

Theorem 2.2.2 *If f is holomorphic on an open set U in $\mathbb{C}$, then $\log|f|$ is subharmonic on U.*

Proof. Evidently $u := \log|f|$ is upper semicontinuous. Also it satisfies the local submean inequality at each $w \in U$ for which $u(w) > -\infty$, because near such a point $\log|f|$ is actually harmonic. On the other hand, if $u(w) = -\infty$, then (2.1) is obvious anyway. $\square$

Further examples can be generated using the following elementary result, which is an immediate consequence of Definition 2.2.1.

Theorem 2.2.3 *Let u and v be subharmonic functions on an open set U in $\mathbb{C}$. Then:*

(a) $\max(u, v)$ *is subharmonic on U;*

(b) $\alpha u + \beta v$ *is subharmonic on U for all $\alpha, \beta \geq 0$.* $\square$

From (a) it follows that subharmonic functions need not be smooth. One might reasonably guess that they do have to be continuous, but actually this is not true either. An example is given in Exercise 2 below, and another in Theorem 2.5.4.

Exercises 2.2

1. Use the Cauchy–Schwarz inequality to show that if h is a harmonic function on an open set U in $\mathbb{C}$, then h^2 is subharmonic on U.

2. Prove that if $\zeta \in \mathbb{C}$ and $r > 0$ then
$$\frac{1}{2\pi} \int_0^{2\pi} \log |re^{it} - \zeta|\, dt = \begin{cases} \log |\zeta|, & \text{if } r \leq |\zeta|, \\ \log r, & \text{if } r > |\zeta|. \end{cases}$$
Use this to show that the function
$$u(z) := \sum_{n \geq 1} 2^{-n} \log |z - 2^{-n}|$$
is subharmonic on $\mathbb{C}$. Show also that u is discontinuous at 0.

2.3 The Maximum Principle

As a consequence of Theorems 1.2.7 and 1.1.6, the local mean-value property implies the (global) mean-value property. To make much further progress with subharmonic functions, we need a corresponding result for the submean inequality. As with harmonic functions, we shall deduce this via a maximum principle. The importance of the maximum principle lies in the fact that from local assumptions it derives global conclusions. Such results are usually very powerful, and the maximum principle is no exception. Since it will feature prominently in what follows, we shall digress slightly in order to study it in a little more detail, returning to the submean inequality in the next section.

Theorem 2.3.1 (Maximum Principle) *Let u be a subharmonic function on a domain D in $\mathbb{C}$.*

(a) *If u attains a global maximum on D, then u is constant.*

(b) *If $\limsup_{z \to \zeta} u(z) \leq 0$ for all $\zeta \in \partial D$, then $u \leq 0$ on D.*

Note that in part (a), u can attain a local maximum or a global minimum without being constant on D. For example, the non-constant subharmonic function $u(z) := \max(\operatorname{Re} z, 0)$ does both on $\mathbb{C}$. Also, just as in Theorem 1.1.8, the validity of part (b) depends on our convention that $\infty \in \partial D$ whenever D is unbounded.

Proof. (a) Suppose that u attains a maximum value M on D. Define
$$A = \{z \in D : u(z) < M\} \quad \text{and} \quad B = \{z \in D : u(z) = M\}.$$
Then A is open because u is upper semicontinuous. Also B is open, because if $u(w) = M$ then the local submean inequality forces u to be equal to M on all sufficiently small circles round w. Clearly A and B partition D so, as D is connected, either $A = D$ or $B = D$. By assumption $B \neq \emptyset$, so $B = D$.

(b) Extend u to ∂D by defining $u(\zeta) = \limsup_{z \to \zeta} u(z)$ ($\zeta \in \partial D$). Then u is upper semicontinuous on $\overline{D}$, which is compact, so by Theorem 2.1.2 u attains a maximum at some $w \in \overline{D}$. If $w \in \partial D$, then by assumption $u(w) \leq 0$, so $u \leq 0$ on D. On the other hand, if $w \in D$, then by part (a) u is constant on D, hence on $\overline{D}$, and so again $u \leq 0$ on D. $\square$

In fact, it *is* possible to replace ∂D by $\partial D \setminus \{\infty\}$ in part (b) if u does not grow too rapidly at infinity. Here is a rather general result that makes this statement precise.

Theorem 2.3.2 (Phragmén–Lindelöf Principle) *Let u be a subharmonic function on an unbounded domain D in $\mathbb{C}$ such that*
$$\limsup_{z \to \zeta} u(z) \leq 0 \quad (\zeta \in \partial D \setminus \{\infty\}).$$
Suppose also that there exists a finite-valued superharmonic function v on D such that
$$\liminf_{z \to \infty} v(z) > 0 \quad \text{and} \quad \limsup_{z \to \infty} \frac{u(z)}{v(z)} \leq 0.$$
Then $u \leq 0$ on D.

Proof. Assume first that $v > 0$ on D. Let $\epsilon > 0$ and set $u_\epsilon = u - \epsilon v$. Then u_ϵ is subharmonic on D, and $\limsup_{z \to \zeta} u_\epsilon(z) \leq 0$ for all $\zeta \in \partial D$ (even ∞), so by Theorem 2.3.1 (b) $u_\epsilon \leq 0$ on D. Letting $\epsilon \to 0$ we get $u \leq 0$ on D.

Now consider a general v. Let $\eta > 0$ and set $F_\eta = \{z \in D : u(z) \geq \eta\}$. Since v is lower semicontinuous and $\liminf_{z \to \infty} v(z) > 0$, it follows that v is bounded below on F_η. Adding a constant to v, if necessary, we can suppose that $v > 0$ there. Set $V = \{z \in D : v(z) > 0\}$. Then for $\zeta \in \partial V \setminus \{\infty\}$ we have
$$\limsup_{z \to \zeta}(u(z) - \eta) \leq \left\{ \begin{array}{ll} \limsup_{z \to \zeta} u(z), & \text{if } \zeta \in \partial D \setminus \{\infty\} \\ u(\zeta) - \eta, & \text{if } \zeta \in D \cap \partial V \end{array} \right\} \leq 0.$$
Applying the special case of the theorem already proved to $u - \eta$ on each component of V, we get $u - \eta \leq 0$ on V. As $F_\eta \subset V$ it follows that $u \leq \eta$ on F_η, and plainly $u \leq \eta$ on $D \setminus F_\eta$, so in fact $u \leq \eta$ on D. Letting $\eta \to 0$ we again deduce that $u \leq 0$ on D. $\square$

2.3. THE MAXIMUM PRINCIPLE

Corollary 2.3.3 *Let u be a subharmonic function on an unbounded proper subdomain D of $\mathbb{C}$ such that*

$$\limsup_{z \to \zeta} u(z) \leq 0 \quad (\zeta \in \partial D \setminus \{\infty\}) \quad \text{and} \quad \limsup_{z \to \infty} \frac{u(z)}{\log |z|} \leq 0.$$

Then $u \leq 0$ on D.

Proof. Take $w \in \partial D$ and apply Theorem 2.3.2 with $v(z) = \log|z-w|$. □

Corollary 2.3.4 (Liouville Theorem) *Let u be a subharmonic function on $\mathbb{C}$ such that*

$$\limsup_{z \to \infty} \frac{u(z)}{\log |z|} \leq 0.$$

Then u is constant on $\mathbb{C}$. In particular, every subharmonic function on $\mathbb{C}$ which is bounded above must be constant.

Proof. If $u \equiv -\infty$ this is clear. Otherwise choose $w \in \mathbb{C}$ with $u(w) > -\infty$, and apply Corollary 2.3.3 to $u - u(w)$ on $\mathbb{C} \setminus \{w\}$. The conclusion is that $u \leq u(w)$ on $\mathbb{C}$. By Theorem 2.3.1 (a) it follows that u is constant. □

For domains of a particular shape one need assume less about growth near infinity. We consider two examples: strips and sectors. These give rise to the classical forms of the Phragmén–Lindelöf principle.

Theorem 2.3.5 *Let S_γ be the strip $\{z \in \mathbb{C} : |\operatorname{Re} z| < \pi/(2\gamma)\}$, where $\gamma > 0$, and let u be a subharmonic function on S_γ such that, for some constants $A < \infty$ and $\alpha < \gamma$,*

$$u(x+iy) \leq A e^{\alpha |y|} \quad (x+iy \in S_\gamma).$$

If $\limsup_{z \to \zeta} u(z) \leq 0$ for all $\zeta \in \partial S_\gamma \setminus \{\infty\}$, then $u \leq 0$ on S_γ.

The function $u(z) := \operatorname{Re}(\cos(\gamma z)) = \cos(\gamma x)\cosh(\gamma y)$ shows that the result is no longer true when $\alpha = \gamma$.

Proof. Choose β with $\alpha < \beta < \gamma$, and define $v \colon S_\gamma \to \mathbb{R}$ by

$$v(z) = \operatorname{Re}(\cos(\beta z)) = \cos(\beta x)\cosh(\beta y) \quad (z = x+iy \in S_\gamma).$$

Then v is harmonic on S_γ,

$$\liminf_{z \to \infty} v(z) \geq \liminf_{|y| \to \infty} \cos(\beta \pi / 2\gamma) \cosh(\beta y) = \infty,$$

and

$$\limsup_{z \to \infty} \frac{u(z)}{v(z)} \leq \limsup_{|y| \to \infty} \frac{A e^{\alpha |y|}}{\cos(\beta \pi / 2\gamma) \cosh(\beta y)} = 0.$$

Hence the result follows from Theorem 2.3.2. □

Corollary 2.3.6 (Three-Lines Theorem) *Let u be a subharmonic function on the strip $S := \{z : 0 < \operatorname{Re} z < 1\}$ such that, for some constants $A < \infty$ and $\alpha < \pi$,*

$$u(x+iy) \leq Ae^{\alpha|y|} \quad (x+iy \in S).$$

If

$$\limsup_{z \to \zeta} u(z) \leq \begin{cases} M_0, & \operatorname{Re}\zeta = 0, \\ M_1, & \operatorname{Re}\zeta = 1, \end{cases}$$

then

$$u(x+iy) \leq M_0(1-x) + M_1 x \quad (x+iy \in S).$$

Proof. Define $\widetilde{u} \colon S \to [-\infty, \infty)$ by

$$\widetilde{u}(z) = u(z) - \operatorname{Re}(M_0(1-z) + M_1 z) \quad (z \in S).$$

Then applying (a translated version of) Theorem 2.3.5 with $\gamma = \pi$ gives $\widetilde{u} \leq 0$ on S. $\square$

Theorem 2.3.7 *Let T_γ be the sector $\{z \in \mathbb{C} \setminus \{0\} : |\arg(z)| < \pi/(2\gamma)\}$, where $\gamma > 1/2$, and let u be a subharmonic function on T_γ such that, for some constants $A, B < \infty$ and $\alpha < \gamma$,*

$$u(z) \leq A + B|z|^\alpha \quad (z \in T_\gamma).$$

If $\limsup_{z \to \zeta} u(z) \leq 0$ for all $\zeta \in \partial T_\gamma \setminus \{\infty\}$, then $u \leq 0$ on T_γ.

Proof. Choose β with $\alpha < \beta < \gamma$, and define $v \colon T_\gamma \to \mathbb{R}$ by

$$v(z) = \operatorname{Re}(z^\beta) = r^\beta \cos(\beta t) \quad (z = re^{it} \in T_\gamma).$$

Then v is harmonic on T_γ,

$$\liminf_{z \to \infty} v(z) \geq \liminf_{r \to \infty} r^\beta \cos(\beta \pi / 2\gamma) = \infty,$$

and

$$\limsup_{z \to \infty} \frac{u(z)}{v(z)} \leq \limsup_{r \to \infty} \frac{A + Br^\alpha}{r^\beta \cos(\beta \pi / 2\gamma)} = 0.$$

Hence again the result follows from Theorem 2.3.2. $\square$

The function $u(z) := \operatorname{Re}(z^\gamma)$ shows that the theorem is no longer true when $\alpha = \gamma$, but we do have the following partial result (in which, for simplicity, we take $\gamma = 1$).

2.3. THE MAXIMUM PRINCIPLE

Corollary 2.3.8 *Let u be a subharmonic function on the half-plane $H := \{z : \operatorname{Re} z > 0\}$ such that, for some constants $A, B < \infty$,*
$$u(z) \le A + B|z| \quad (z \in H).$$
If $\limsup_{z \to \zeta} u(z) \le 0$ for all $\zeta \in \partial H \setminus \{\infty\}$, and $\limsup_{x \to \infty} u(x)/x = L$, then
$$u(z) \le L(\operatorname{Re} z) \quad (z \in H).$$

Proof. Given $L' > L$, define $\tilde{u} \colon H \to [-\infty, \infty)$ by
$$\tilde{u}(z) = u(z) - L'(\operatorname{Re} z) \quad (z \in H).$$
Then applying (a rotated version of) Theorem 2.3.7 with $\gamma = 2$ on each of the two sectors $-\pi/2 < \arg(z) < 0$ and $0 < \arg(z) < \pi/2$, we deduce that $\tilde{u}$ is bounded above on H. Applying Theorem 2.3.7 again, this time with $\gamma = 1$, gives that $\tilde{u} \le 0$ on H. Hence $u(z) \le L'(\operatorname{Re} z)$, and as this is true for each $L' > L$, the result follows. □

Exercises 2.3

1. Let $u_1, \ldots, u_n$ be subharmonic functions on a domain D in $\mathbb{C}$, and suppose that their sum $u_1 + \cdots + u_n$ attains a maximum on D. Show that each function u_j is harmonic on D.

2. Let u be a subharmonic function on $\Delta(0, 1)$ such that $u \le 0$. Prove that for each $\zeta \in \partial \Delta(0, 1)$,
$$\limsup_{r \to 1^-} \frac{u(r\zeta)}{1 - r} < 0.$$
[Hint: apply the maximum principle to $u(z) + c \log|z|$ on the set $\{1/2 < |z| < 1\}$ for a suitable constant c.]

3. Let $\Delta = \Delta(0, 1)$ and $f \colon \Delta \to \Delta$ be a holomorphic function such that
$$f(z) = z + o(|1 - z|^3) \quad \text{as } z \to 1.$$
   (i) Let $\phi(z) = (1 + z)/(1 - z)$ and $u(z) = \operatorname{Re}(\phi(z) - \phi(f(z)))$. Show that $\limsup_{z \to \zeta} u(z) \le 0$ for each $\zeta \in \partial \Delta \setminus \{1\}$, and that $u(z) = o(|1 - z|)$ as $z \to 1$.
   (ii) Use the maximum principle to deduce that $u \le 0$, and then Exercise 2 to show that $u \equiv 0$.
   (iii) Conclude that $f(z) \equiv z$.

   Give an example to show that the conclusion in (iii) fails if we merely suppose that $f(z) = z + O(|1 - z|^3)$ as $z \to 1$.

4. Let u be a subharmonic function on $\Delta(0,1)$ such that
$$u(z) \leq -\log|\operatorname{Im} z| \quad (|z| < 1).$$
Prove that
$$u(z) \leq -\log\left|\frac{1-z^2}{2}\right| \quad (|z| < 1).$$
[Hint: apply the maximum principle to $u(z) + \log|(r^2 - z^2)/2r|$ on $\Delta(0,r)$, where $r < 1$, and then let $r \to 1$.]

5. Let u be a subharmonic function on $H := \{z : \operatorname{Re} z > 0\}$, and suppose that there exist constants $A, B < \infty$ and $\alpha > 0$ such that
$$(2.2) \quad \begin{cases} u(z) \leq A + B|z| & (z \in H), \\ \limsup_{z \to \zeta} u(z) \leq -\alpha|\zeta| & (\zeta \in \partial H \setminus \{\infty\}). \end{cases}$$

(i) Apply Theorem 2.3.7 to $u(z) - A - B(\operatorname{Re} z) + \alpha(\operatorname{Im} z)$ on the sector $0 < \arg(z) < \pi/2$ to show that
$$u(z) \leq A + B(\operatorname{Re} z) - \alpha(\operatorname{Im} z) \quad (0 < \arg(z) < \pi/2).$$

(ii) Deduce that u is bounded above on the line $\arg(z) = \theta$, where $\tan \theta = B/\alpha$.

(iii) Apply Theorem 2.3.7 on the sectors $-\pi/2 < \arg(z) < \theta$ and $\theta < \arg(z) < \pi/2$ to show that u is bounded above on H.

(iv) Apply Theorem 2.3.7 on H to show that $u \leq 0$ on H.

(v) Show that if $M > 0$ then $\tilde{u} := u + M(\operatorname{Re} z)$ also satisfies (2.2) (with different constants), and deduce that $u \leq -M(\operatorname{Re} z)$ on H.

(vi) Conclude that $u \equiv -\infty$ on H.

6. Let f be a holomorphic function on $H := \{z : \operatorname{Re} z > 0\}$ such that, for some constants $C < \infty$ and $\gamma < \pi$,
$$|f(z)| \leq Ce^{\gamma|z|} \quad (z \in H).$$
Suppose also that $f(n) = 0$ for each integer $n \geq 1$. Show that
$$u(z) := \log\left|\frac{f(z)}{2C\sin(\pi z)}\right|$$
is subharmonic on H, and that it satisfies (2.2) with $\alpha = \pi - \gamma > 0$. Deduce that $f \equiv 0$ on H.

2.4 Criteria for Subharmonicity

Now that the necessary tools are available, we can prove that subharmonic functions satisfy the global submean inequality. In fact more is true: they also obey an inequality corresponding to the Poisson integral formula, as is shown by the following theorem.

Theorem 2.4.1 *Let U be an open subset of $\mathbb{C}$, and let $u: U \to [-\infty, \infty)$ be an upper semicontinuous function. Then the following are equivalent.*

(a) *The function u is subharmonic on U.*

(b) *Whenever $\overline{\Delta}(w, \rho) \subset U$, then for $r < \rho$ and $0 \leq t < 2\pi$*

$$u(w + re^{it}) \leq \frac{1}{2\pi} \int_0^{2\pi} \frac{\rho^2 - r^2}{\rho^2 - 2\rho r \cos(\theta - t) + r^2} u(w + \rho e^{i\theta}) \, d\theta.$$

(c) *Whenever D is a relatively compact subdomain of U, and h is a harmonic function on D satisfying*

$$\limsup_{z \to \zeta}(u - h)(z) \leq 0 \quad (\zeta \in \partial D),$$

then $u \leq h$ on D.

Proof. (a)$\Rightarrow$(c): Given D and h as in (c), the function $u - h$ is subharmonic on D, so the result follows by the maximum principle, Theorem 2.3.1 (b).

(c)$\Rightarrow$(b): Suppose that $\overline{\Delta} := \overline{\Delta}(w, \rho) \subset U$. By Theorem 2.1.3 there exist continuous functions $\phi_n: \partial \Delta \to \mathbb{R}$ such that $\phi_n \downarrow u$ on $\partial \Delta$. By Theorem 1.2.4 each $P_\Delta \phi_n$ is harmonic on Δ. Also $\lim_{z \to \zeta} P_\Delta \phi_n(z) = \phi_n(\zeta)$ for all $\zeta \in \partial \Delta$, and hence

$$\limsup_{z \to \zeta}(u - P_\Delta \phi_n)(z) \leq u(\zeta) - \phi_n(\zeta) \leq 0 \quad (\zeta \in \partial \Delta).$$

From (c) it follows that $u \leq P_\Delta \phi_n$ on Δ. Letting $n \to \infty$ and using the monotone convergence theorem gives the desired inequality.

(b)$\Rightarrow$(a): This is clear. $\square$

Corollary 2.4.2 (Global Submean Inequality) *If u is a subharmonic function on an open set U in $\mathbb{C}$, and if $\overline{\Delta}(w, \rho) \subset U$, then*

$$u(w) \leq \frac{1}{2\pi} \int_0^{2\pi} u(w + \rho e^{i\theta}) \, d\theta.$$

Proof. Put $r = 0$ in Theorem 2.4.1 (b). $\square$

The criterion (c) in Theorem 2.4.1, as well as explaining the name 'subharmonic', is also useful in its own right. For example, since it remains invariant under conformal mapping, we immediately deduce the following corollary.

Corollary 2.4.3 *If $f: U_1 \to U_2$ is a conformal mapping between open subsets U_1 and U_2 of $\mathbb{C}$, and if u is subharmonic on U_2, then $u \circ f$ is subharmonic on U_1.* □

Using this result, we can extend the definition of subharmonicity to the Riemann sphere in just the same way as we did for harmonicity after Corollary 1.1.5. It is easily checked that all the results of Section 2.2 remain valid for subharmonic functions defined on open subsets of $\mathbb{C}_\infty$, as does the maximum principle, Theorem 2.3.1.

Corollary 2.4.3 remains true for a general holomorphic function f. One proof is outlined in Exercise 2, and another will be given in Section 2.7.

As another application of Theorem 2.4.1, we can characterize those C^2 functions which are subharmonic as those with positive Laplacian. This result vindicates what was said at the beginning of Section 2.2.

Theorem 2.4.4 *Let U be an open subset of $\mathbb{C}$, and let $u \in C^2(U)$. Then u is subharmonic on U if and only if $\Delta u \geq 0$ on U.*

Proof. Assume first that $\Delta u \geq 0$ on U. We shall use Theorem 2.4.1 (c) to prove that u is subharmonic. Let D be a relatively compact subdomain of U, and suppose that h is a harmonic function on D such that $\limsup_{z \to \zeta}(u-h)(z) \leq 0$ for all $\zeta \in \partial D$. We need to show that $u \leq h$ on D. Let $\epsilon > 0$ and define

$$v_\epsilon(z) = \begin{cases} u(z) - h(z) + \epsilon|z|^2, & \text{if } z \in D, \\ \epsilon|z|^2, & \text{if } z \in \partial D. \end{cases}$$

Then v_ϵ is upper semicontinuous on $\overline{D}$, so it attains a maximum there. But v_ϵ cannot attain a local maximum on D because $\Delta v_\epsilon = \Delta u + 4\epsilon > 0$ on D. Therefore the maximum is attained on ∂D, and hence $u - h \leq \sup_{\partial D} \epsilon|z|^2$ on D. Letting $\epsilon \to 0$ yields the desired conclusion.

Conversely, assume that u is subharmonic on U. Suppose, if possible, that $\Delta u(w) < 0$ for some $w \in U$. Then by continuity, there exists $\rho > 0$ such that $\Delta u \leq 0$ on $\Delta(w, \rho)$. By what we have just proved, this implies that u is superharmonic on $\Delta(w, \rho)$, and hence harmonic there. In particular $\Delta u(w) = 0$, which contradicts the original supposition. Therefore $\Delta u \geq 0$ on U. □

2.4. CRITERIA FOR SUBHARMONICITY

The next result, which nicely illustrates the flexibility of subharmonic functions, shows how they can be 'glued together'.

Theorem 2.4.5 (Gluing Theorem) *Let u be a subharmonic function on an open set U in $\mathbb{C}$, and let v be a subharmonic function on an open subset V of U such that*

$$\limsup_{z \to \zeta} v(z) \leq u(\zeta) \quad (\zeta \in U \cap \partial V).$$

Then $\widetilde{u}$ is subharmonic on U, where

$$\widetilde{u} = \begin{cases} \max(u, v) & \text{on } V, \\ u & \text{on } U \setminus V. \end{cases}$$

Proof. The boundary condition on v ensures that $\widetilde{u}$ is upper semicontinuous on U. By Theorem 2.2.3 (a) $\widetilde{u}$ satisfies the local submean inequality at each $w \in V$, and it also does so when $w \in U \setminus V$ because $\widetilde{u} \geq u$ on U. □

We conclude this section with three theorems about infinite families of subharmonic functions. The first of these, for decreasing sequences, is simple but important. It would no longer be true if we were to restrict subharmonic functions to be continuous, and indeed is one of the principal reasons for not doing so.

Theorem 2.4.6 *Let $(u_n)_{n \geq 1}$ be subharmonic functions on an open set U in $\mathbb{C}$, and suppose that $u_1 \geq u_2 \geq u_3 \geq \cdots$ on U. Then $u := \lim_{n \to \infty} u_n$ is subharmonic on U.*

Proof. For each $\alpha \in \mathbb{R}$, the set $\{z : u(z) < \alpha\}$ is the union of the open sets $\{z : u_n(z) < \alpha\}$, so it is open and u is upper semicontinuous.

Also, if $\overline{\Delta}(w, \rho) \subset U$, then for each $n \geq 1$

$$u_n(w) \leq \frac{1}{2\pi} \int_0^{2\pi} u_n(w + \rho e^{i\theta}) \, d\theta.$$

Letting $n \to \infty$ and applying the monotone convergence theorem we deduce that u satisfies the submean inequality. □

The corresponding result for an increasing sequence (u_n) is false because, even if it is finite, the limit u may fail to be upper semicontinuous. For example, if $u_n(z) = (1/n) \log |z|$ on $\Delta(0, 1)$, then

$$u(z) = \begin{cases} 0, & \text{if } 0 < |z| < 1, \\ -\infty, & \text{if } z = 0. \end{cases}$$

We shall return to this topic in Section 3.4.

The remaining two results generalize Theorem 2.2.3 (a) and (b) respectively.

Theorem 2.4.7 *Let T be a compact topological space, let U be an open subset of $\mathbb{C}$, and let $v: U \times T \to [-\infty, \infty)$ be a function such that:*

(a) *v is upper semicontinuous on $U \times T$;*

(b) *$z \mapsto v(z,t)$ is subharmonic on U for each $t \in T$.*

Then $u(z) := \sup_{t \in T} v(z,t)$ is subharmonic on U.

Proof. Let $w \in U$ and suppose that $u(w) < \alpha$. Then for each $t \in T$ we have $v(w,t) < \alpha$, so as v is upper semicontinuous, there exists a neighbourhood N_t of t and $\rho_t > 0$ such that $v < \alpha$ on $\Delta(w, \rho_t) \times N_t$. As T is compact, it has a finite subcover $N_{t_1}, \ldots, N_{t_n}$. Then $u < \alpha$ on $\Delta(w, \rho')$, where $\rho' = \min(\rho_{t_1}, \ldots, \rho_{t_n})$. This shows that u is upper semicontinuous.

Now suppose that $\overline{\Delta}(w, \rho) \subset U$. Then for each $t \in T$,

$$v(w,t) \leq \frac{1}{2\pi} \int_0^{2\pi} v(w + \rho e^{i\theta}, t) \, d\theta \leq \frac{1}{2\pi} \int_0^{2\pi} u(w + \rho e^{i\theta}) \, d\theta.$$

Taking the supremum over $t \in T$, we deduce that u satisfies the submean inequality. □

Theorem 2.4.8 *Let (Ω, μ) be a measure space with $\mu(\Omega) < \infty$, let U be an open subset of $\mathbb{C}$, and let $v: U \times \Omega \to [-\infty, \infty)$ be a function such that:*

(a) *v is measurable on $U \times \Omega$;*

(b) *$z \mapsto v(z, \omega)$ is subharmonic on U for each $\omega \in \Omega$;*

(c) *$z \mapsto \sup_{\omega \in \Omega} v(z, \omega)$ is locally bounded above on U.*

Then $u(z) := \int_\Omega v(z, \omega) \, d\mu(\omega)$ is subharmonic on U.

Proof. It is sufficient to prove that u is subharmonic on each relatively compact subdomain D of U. Fix such a D. Then (c) implies that $\sup_\omega v(z, \omega)$ is bounded above on D, so by subtracting a constant, if necessary, we can suppose that $v \leq 0$ on $D \times \Omega$. This legitimizes the use of Fatou's lemma and Fubini's theorem in what follows.

Whenever $w_n \to w$ in D, then by Fatou's lemma

$$\begin{aligned} \limsup_{n \to \infty} u(w_n) &\leq \int_\Omega \limsup_{n \to \infty} v(w_n, \omega) \, d\mu(\omega) \\ &\leq \int_\Omega v(w, \omega) \, d\mu(\omega) = u(w). \end{aligned}$$

It follows that u is upper semicontinuous on D.

2.5. INTEGRABILITY

Also, if $\overline{\Delta}(w,\rho) \subset D$, then by Fubini's theorem

$$\frac{1}{2\pi}\int_0^{2\pi} u(w+\rho e^{i\theta})\,d\theta = \int_\Omega \left(\frac{1}{2\pi}\int_0^{2\pi} v(w+\rho e^{i\theta},\omega)\,d\theta\right) d\mu(\omega)$$
$$\geq \int_\Omega v(w,\omega)\,d\mu(\omega) = u(w),$$

so that u satisfies the submean inequality on D. $\square$

Exercises 2.4

1. Let u be an upper semicontinuous function on an open subset U of $\mathbb{C}$, and suppose that for each $w \in U$ with $u(w) > -\infty$ we have

$$\limsup_{r\to 0}\frac{1}{r^2}\left(\frac{1}{2\pi}\int_0^{2\pi} u(w+re^{it})\,dt - u(w)\right) \geq 0.$$

   Prove that u is subharmonic on U. [Hint: Let $\epsilon > 0$ and set $u_\epsilon = u + \epsilon|z|^2$. Repeat the steps leading up to Corollary 2.4.2 to show that u_ϵ satisfies the submean inequality. Then let $\epsilon \to 0$.]

2. (i) Show that if $u(z)$ is subharmonic on a neighbourhood of 0, then so is $u(z^k)$ for each $k \geq 1$.
   (ii) Show that if f is holomorphic on a neighbourhood of w and $f - f(w)$ vanishes to order exactly k at w, then there exists an injective holomorphic function g on a neighbourhood of w such that $f(z) = g(z)^k + f(w)$ there.
   (iii) Combine (i) and (ii) to extend Corollary 2.4.3 to arbitrary holomorphic functions.

3. Let U be an open subset of $\mathbb{C}$. Show that $u(z) := -\log \text{dist}(z, \partial U)$ is subharmonic on U.

2.5 Integrability

As a subharmonic function is upper semicontinuous, it is automatically bounded above on compact sets (Theorem 2.1.2). More subtle is the fact that also it cannot be 'too unbounded below'.

Theorem 2.5.1 (Integrability Theorem) *Let u be a subharmonic function on a domain D in $\mathbb{C}$, with $u \not\equiv -\infty$ on D. Then u is locally integrable on D, i.e. $\int_K |u|\,dA < \infty$ for each compact subset K of D.*

(Here, and throughout, dA denotes two-dimensional Lebesgue measure.)

Proof. By a simple compactness argument, it suffices to show that for each $w \in D$ there exists $\rho > 0$ such that

$$\int_{\Delta(w,\rho)} |u|\, dA < \infty. \tag{2.3}$$

Let A denote the set of $w \in D$ which have this property, and let B be the set of those that do not. We shall show that both A and B are open, and that $u = -\infty$ on B, from which the result follows by connectedness of D.

Let $w \in A$, and choose $\rho > 0$ such that (2.3) holds. Given $w' \in \Delta(w,\rho)$, set $\rho' = \rho - |w' - w|$. Then $\Delta(w',\rho') \subset \Delta(w,\rho)$, so

$$\int_{\Delta(w',\rho')} |u|\, dA < \infty.$$

Therefore $\Delta(w,\rho) \subset A$, which shows that A is open.

Now let $w \in B$, and choose $\rho > 0$ so that $\overline{\Delta}(w,2\rho) \subset D$. Then because $w \in B$,

$$\int_{\Delta(w,\rho)} |u|\, dA = \infty.$$

Given $w' \in \Delta(w,\rho)$, set $\rho' = \rho + |w' - w|$. Then $\Delta(w',\rho') \supset \Delta(w,\rho)$ and u is bounded above on $\overline{\Delta}(w',\rho')$, so

$$\int_{\Delta(w',\rho')} u\, dA = -\infty.$$

Now u satisfies the submean inequality

$$u(w') \leq \frac{1}{2\pi} \int_0^{2\pi} u(w' + re^{it})\, dt \quad (0 \leq r \leq \rho'),$$

so multiplying by $2\pi r$ and integrating from $r = 0$ to $r = \rho'$ gives

$$\pi \rho'^2 u(w') \leq \int_{\Delta(w',\rho')} u\, dA = -\infty.$$

Hence $u = -\infty$ on $\Delta(w,\rho)$. This shows that B is open and that $u = -\infty$ on B, as required. □

From this, it follows that subharmonic functions are also integrable on circles.

Corollary 2.5.2 *Let u be a subharmonic function on a domain D in $\mathbb{C}$, with $u \not\equiv -\infty$. If $\overline{\Delta}(w,\rho) \subset D$, then*

$$\frac{1}{2\pi} \int_0^{2\pi} u(w + \rho e^{i\theta})\, d\theta > -\infty.$$

2.5. INTEGRABILITY

Proof. Fix $\overline{\Delta}(w,\rho) \subset D$. Since u is bounded above on compact sets, by subtracting a constant we can suppose that $u \leq 0$ on $\overline{\Delta}(w,\rho)$. By Theorem 2.4.1 (b), if $r < \rho$ and $0 \leq t < 2\pi$ then

$$u(w+re^{it}) \leq \frac{1}{2\pi} \int_0^{2\pi} \frac{\rho^2 - r^2}{\rho^2 - 2\rho r \cos(\theta - t) + r^2} u(w + \rho e^{i\theta})\, d\theta$$

$$\leq \left(\frac{\rho - r}{\rho + r}\right) \frac{1}{2\pi} \int_0^{2\pi} u(w + \rho e^{i\theta})\, d\theta.$$

Hence, if the last integral were $-\infty$, it would imply that $u \equiv -\infty$ on $\Delta(w,\rho)$, contradicting Theorem 2.5.1. Therefore the integral is finite. □

Another consequence of Theorem 2.5.1 is that subharmonic functions can only be equal to $-\infty$ on relatively small sets.

Corollary 2.5.3 *Let u be a subharmonic function on a domain D in $\mathbb{C}$, with $u \not\equiv -\infty$ on D. Then*

$$E := \{z \in D : u(z) = -\infty\}$$

is a set of Lebesgue measure zero.

Proof. Let $(K_n)_{n \geq 1}$ be compact sets with $\cup_n K_n = D$. For each n we have $\int_{K_n} |u|\, dA < \infty$, so $E \cap K_n$ has measure zero. Since $E = \cup_n (E \cap K_n)$, it too has measure zero. □

The set E above is also small in other ways: one is outlined in Exercise 1, and others will occur in Chapter 3. Of course, if $u = \log|f|$ where f is holomorphic, then E is just the zero set of f, and is therefore countable. But as the following construction shows, there are subharmonic functions which are $-\infty$ on uncountable sets.

Theorem 2.5.4 *Let K be a compact subset of $\mathbb{C}$ with no isolated points, let $(w_n)_{n \geq 1}$ be a countable dense subset of K, and let $(a_n)_{n \geq 1}$ be strictly positive numbers such that $\sum_n a_n < \infty$. Define $u: \mathbb{C} \to [-\infty, \infty)$ by*

$$u(z) = \sum_{n \geq 1} a_n \log|z - w_n| \quad (z \in \mathbb{C}).$$

Then:

(a) *u is subharmonic on $\mathbb{C}$ and $u \not\equiv -\infty$;*

(b) *$u = -\infty$ on an uncountable dense subset of K;*

(c) *u is discontinuous almost everywhere on K.*

Proof. (a) Let μ be the finite measure on $\mathbb{N}$ given by $\mu(\{n\}) = a_n$ ($n \geq 1$), and define $v\colon \mathbb{C} \times \mathbb{N} \to [-\infty, \infty)$ by $v(z,n) = \log|z - w_n|$. Then

$$\int_{\mathbb{N}} v(z,n)\, d\mu(n) = \sum_{n \geq 1} a_n \log|z - w_n| = u(z) \quad (z \in \mathbb{C}),$$

so by Theorem 2.4.8 u is subharmonic on $\mathbb{C}$. Also $u(z) > -\infty$ for all $z \in \mathbb{C} \setminus K$, so $u \not\equiv -\infty$.

(b) Set $E = \{z \in \mathbb{C} : u(z) = -\infty\}$. Clearly $E \subset K$, and each $w_n \in E$ so E is dense in K. Since

$$K \setminus E = \bigcup_{n \geq 1} \{z \in K : u(z) \geq -n\},$$

a countable union of closed nowhere dense sets, it follows that $K \setminus E$ is meagre in K. If E were countable, it would imply that K was meagre in itself, contradicting the Baire category theorem. Therefore E is uncountable.

(c) The function u is discontinuous at every point of $\overline{E} \setminus E$. Since E is dense in K, and by Corollary 2.5.3 E has measure zero, it follows that u is discontinuous almost everywhere on K. $\square$

We shall return to study the sets where subharmonic functions are $-\infty$ in more detail in Section 3.5.

Exercises 2.5

1. Let u be a subharmonic function on a domain D in $\mathbb{C}$, and suppose that $u = -\infty$ on a straight line segment L in D.

    (i) Let Δ be an open disc in D such that L contains a diameter of Δ, splitting Δ into Δ^+ and Δ^- say. Define $v\colon \Delta \to [-\infty, \infty)$ by

    $$v(z) = \begin{cases} u(z), & \text{if } z \in \Delta^+, \\ -\infty, & \text{if } z \in \Delta^- \cup L. \end{cases}$$

    Show that v is subharmonic on Δ.

    (ii) Show that $v \equiv -\infty$ on Δ, and deduce that $u = -\infty$ on Δ^+.

    (iii) Conclude that $u \equiv -\infty$ on D.

2. Is it possible for a subharmonic function on a domain D to be discontinuous at *every* point of D? [Hint: look at Exercise 2.1.3.]

2.6 Convexity

As we have already remarked, there are strong similarities between subharmonic functions on $\mathbb{C}$ and convex functions on $\mathbb{R}$. In this section we examine in more detail the relationship between the two classes.

Definition 2.6.1 Let $-\infty \leq a < b \leq \infty$. A function $\psi \colon (a,b) \to \mathbb{R}$ is called *convex* if, whenever $t_1, t_2 \in (a,b)$,
$$\psi((1-\lambda)t_1 + \lambda t_2) \leq (1-\lambda)\psi(t_1) + \lambda\psi(t_2) \quad (0 \leq \lambda \leq 1).$$

It is well known that convex functions are continuous. Also, given $\psi \in C^2(a,b)$, then ψ is convex if and only if $\psi'' \geq 0$ on (a,b). We shall need a basic inequality for convex functions.

Theorem 2.6.2 (Jensen's Inequality) *Let $-\infty \leq a < b \leq \infty$, and let $\psi \colon (a,b) \to \mathbb{R}$ be a convex function. Also, let (Ω, μ) be a measure space with $\mu(\Omega) = 1$, and let $f \colon \Omega \to (a,b)$ be an integrable function. Then*
$$\psi\left(\int_\Omega f\, d\mu\right) \leq \int_\Omega \psi \circ f\, d\mu.$$

Proof. Set $c = \int_\Omega f\, d\mu$, so $c \in (a,b)$. By convexity, if $a < t_1 < c < t_2 < b$ then
$$\psi(c) \leq \frac{t_2 - c}{t_2 - t_1}\psi(t_1) + \frac{c - t_1}{t_2 - t_1}\psi(t_2).$$
After rearrangement, this implies that
$$\sup_{t_1 \in (a,c)} \frac{\psi(c) - \psi(t_1)}{c - t_1} \leq \inf_{t_2 \in (c,b)} \frac{\psi(t_2) - \psi(c)}{t_2 - c}.$$
Hence there exists a constant M such that
$$\psi(t) \geq \psi(c) + M(t - c) \quad (t \in (a,b)).$$
Putting $t = f(\omega)$ and integrating with respect to μ, we get
$$\int_\Omega \psi(f(\omega))\, d\mu(\omega) \geq \int_\Omega \psi(c)\, d\mu(\omega) + M\int_\Omega (f(\omega) - c)\, d\mu(\omega) = \psi(c),$$
which is the desired inequality. $\square$

This enables us to generate new examples of subharmonic functions.

Theorem 2.6.3 *Let $-\infty \leq a < b \leq \infty$, let $u \colon U \to [a,b)$ be a subharmonic function on an open set U in $\mathbb{C}$, and let $\psi \colon (a,b) \to \mathbb{R}$ be an increasing convex function. Then $\psi \circ u$ is subharmonic on U, where we define $\psi(a) = \lim_{t \to a} \psi(t)$.*

Proof. Choose $(a_n)_{n\geq 1} \in (a,b)$ with $a_n \downarrow a$, and for each n set $u_n = \max(u, a_n)$, so u_n is subharmonic. Then certainly $\psi \circ u_n$ is upper semicontinuous on U. Also if $\overline{\Delta}(w, \rho) \subset U$ then

$$\psi \circ u_n(w) \leq \psi\left(\frac{1}{2\pi} \int_0^{2\pi} u_n(w + \rho e^{i\theta}) \, d\theta\right) \leq \frac{1}{2\pi} \int_0^{2\pi} \psi \circ u_n(w + \rho e^{i\theta}) \, d\theta,$$

the second inequality coming from Jensen's inequality applied to the measure $d\theta/2\pi$ on $[0, 2\pi)$. Hence $\psi \circ u_n$ is subharmonic on U. Since $\psi \circ u_n \downarrow \psi \circ u$ as $n \to \infty$, it follows from Theorem 2.4.6 that $\psi \circ u$ is subharmonic on U. □

Corollary 2.6.4 *If u is subharmonic on an open subset U of $\mathbb{C}$, then so is $\exp u$.* □

For example, applying this result to $u := \alpha \log |f|$, where f is holomorphic and $\alpha > 0$, we deduce that $|f|^\alpha$ is subharmonic.

It is sometimes also of interest to know under what conditions $\log u$ is subharmonic.

Theorem 2.6.5 *Let $u: U \to [0, \infty)$ be a function on an open set U in $\mathbb{C}$. Then $\log u$ is subharmonic on U if and only if $u|e^q|$ is subharmonic on U for every (complex) polynomial q.*

Proof. Suppose first that $\log u$ is subharmonic. Then so is $\log u + \operatorname{Re} q$ for each polynomial q, and taking exponentials, Corollary 2.6.4 implies that $u|e^q|$ is subharmonic.

Conversely, suppose that $u|e^q|$ is subharmonic on U for each polynomial q. Taking $q = 0$, we see straightaway that u is subharmonic, and in particular upper semicontinuous. Hence $\log u$ is also upper semicontinuous. It remains to check the submean inequality. Let $\Delta := \Delta(w, \rho)$ be a disc with $\overline{\Delta} \subset U$, and choose continuous functions $\phi_n: \partial \Delta \to \mathbb{R}$ such that $\phi_n \downarrow \log u$ on $\partial \Delta$. For each $n \geq 1$ we can find a polynomial q_n such that

$$0 \leq \operatorname{Re} q_n - \phi_n \leq 1/n \quad \text{on } \partial \Delta.$$

(This follows easily from the Stone–Weierstrass theorem, or alternatively from Exercise 1.2.2.) Then we have

$$\limsup_{z \to \zeta} u(z)|e^{-q_n(z)}| \leq e^{\phi_n(\zeta)} e^{-\operatorname{Re} q_n(\zeta)} \leq 1 \quad (\zeta \in \partial \Delta).$$

2.6. CONVEXITY

Since $u|e^{-q_n}|$ is assumed subharmonic, it follows from the maximum principle that $u|e^{-q_n}| \leq 1$ on Δ. Hence

$$\log u(w) \leq \operatorname{Re} q_n(w) = \frac{1}{2\pi} \int_0^{2\pi} \operatorname{Re} q_n(w + \rho e^{i\theta}) \, d\theta$$

$$\leq \frac{1}{2\pi} \int_0^{2\pi} \phi_n(w + \rho e^{i\theta}) \, d\theta + \frac{1}{n}.$$

Letting $n \to \infty$ and applying the monotone convergence theorem, we deduce that

$$\log u(w) \leq \frac{1}{2\pi} \int_0^{2\pi} \log u(w + \rho e^{i\theta}) \, d\theta,$$

so $\log u$ does indeed obey the submean inequality. $\square$

Theorem 2.6.3 also allows us to characterize radial subharmonic functions.

Theorem 2.6.6 *Let $v \colon \Delta(0, \rho) \to [-\infty, \infty)$ be a function which is radial (i.e. $v(z) = v(|z|)$ for all z), and assume that $v \not\equiv -\infty$. Then v is subharmonic on $\Delta(0, \rho)$ if and only if $v(r)$ is an increasing convex function of $\log r$ $(0 < r < \rho)$ with $\lim_{r \to 0} v(r) = v(0)$.*

Proof. The 'if' follows by applying Theorem 2.6.3 with $u(z) = \log|z|$ and $\psi(t) = v(e^t)$.

For the 'only if', assume that v is subharmonic on $\Delta(0, \rho)$. Given $r_1, r_2 \in [0, \rho)$ with $r_1 < r_2$, the maximum principle applied to v on $\Delta(0, r_2)$ yields

$$v(r_1) \leq \sup_{\partial \Delta(0, r_2)} v = v(r_2).$$

Hence v is increasing on $[0, \rho)$. Also, it follows that $\liminf_{r \to 0} v(r) \geq v(0)$. On the other hand, upper semicontinuity implies $\limsup_{r \to 0} v(r) \leq v(0)$, and hence $\lim_{r \to 0} v(r) = v(0)$.

It remains to show that $v(r)$ is a convex function of $\log r$. Notice first that by Corollary 2.5.2 $v(r) > -\infty$ for $r > 0$. Given $r_1, r_2 \in (0, \rho)$ with $r_1 < r_2$, choose constants α, β such that $\alpha + \beta \log r = v(r)$ for $r = r_1, r_2$. Applying the maximum principle to $v(z) - \alpha - \beta \log|z|$ on $\{z : r_1 < |z| < r_2\}$, we get

$$v(r) \leq \alpha + \beta \log r \quad (r_1 < r < r_2).$$

Hence if $0 \leq \lambda \leq 1$ and $\log r = (1 - \lambda) \log r_1 + \lambda \log r_2$, then

$$\begin{aligned} v(r) &\leq \alpha + \beta \log r \\ &= (1 - \lambda)(\alpha + \beta \log r_1) + \lambda(\alpha + \beta \log r_2) \\ &= (1 - \lambda) v(r_1) + \lambda v(r_2), \end{aligned}$$

which proves the convexity. $\square$

Theorem 2.6.6 can be used to study various integral means of subharmonic functions.

Definition 2.6.7 Let u be a subharmonic function on the disc $\Delta(0, \rho)$, with $u \not\equiv -\infty$. For $0 < r < \rho$, we define

$$M_u(r) := \sup_{|z|=r} u(z),$$

$$C_u(r) := \frac{1}{2\pi} \int_0^{2\pi} u(re^{it})\, dt,$$

$$B_u(r) := \frac{1}{\pi r^2} \int_{\Delta(0,r)} u\, dA.$$

Note that by Theorem 2.5.1 and Corollary 2.5.2, $M_u(r), C_u(r), B_u(r)$ are all finite. Also $C_u(r)$ and $B_u(r)$ are connected by the relation

(2.4) $$B_u(r) = \frac{2}{r^2} \int_0^r C_u(s) s\, ds.$$

Theorem 2.6.8 *With the notation of Definition 2.6.7:*

(a) $M_u(r), C_u(r), B_u(r)$ are all increasing convex functions of $\log r$;

(b) $M_u(r) \geq C_u(r) \geq B_u(r) \geq u(0)$ $(0 < r < \rho)$;

(c) $\lim_{r \to 0} M_u(r) = \lim_{r \to 0} C_u(r) = \lim_{r \to 0} B_u(r) = u(0)$.

Proof. (a) Observe that for $0 < r < \rho$,

$$M_u(r) = v(r) \qquad \text{where} \qquad v(z) = \sup_{t \in [0, 2\pi]} u(ze^{it}),$$

$$C_u(r) = v(r) \qquad \text{where} \qquad v(z) = \frac{1}{2\pi} \int_0^{2\pi} u(ze^{it})\, dt,$$

$$B_u(r) = v(r) \qquad \text{where} \qquad v(z) = \frac{1}{\pi} \int_0^{2\pi} \int_0^1 u(zse^{it}) s\, ds\, dt.$$

In each case v is subharmonic on $\Delta(0, \rho)$: this is proved using Theorem 2.4.7 in the first case, and Theorem 2.4.8 in the other two. Clearly each v is also radial, and so the result follows from Theorem 2.6.6.

(b) The first inequality is clear. To derive the others, we begin with the relation $C_u(r) \geq C_u(s) \geq u(0)$ $(r \geq s)$, proved in (a). Multiplying both sides by $2s/r^2$ and integrating from $s = 0$ to $s = r$, we get

$$C_u(r) \geq \frac{2}{r^2} \int_0^{2\pi} C_u(s) s\, ds \geq u(0).$$

Combining this with (2.4) gives $C_u(r) \geq B_u(r) \geq u(0)$, as desired.

2.6. CONVEXITY

(c) By (b), it is enough to show that $\limsup_{r \to 0} M_u(r) \leq u(0)$, and this is an immediate consequence of the upper semicontinuity of u. □

These integral means enjoy several other interesting properties, a few of which are outlined in the exercises below.

Exercises 2.6

1. Let $u: \Delta(0, \rho) \to \mathbb{R}$ be a function such that $u(x + iy)$ is convex in x for each fixed y, and convex in y for each fixed x. Prove that u is subharmonic on $\Delta(0, \rho)$. Give an example to show that the converse is false.

2. Let $-\infty \leq a < b \leq \infty$, let $h: U \to (a, b)$ be a harmonic function on an open set U in $\mathbb{C}$, and let $\psi: (a, b) \to \mathbb{R}$ be a convex function (not necessarily increasing). Show that $\psi \circ h$ is subharmonic on U.

3. Let $u: U \to [0, \infty)$ be a function on an open set U in $\mathbb{C}$. Prove that $\log u$ is subharmonic on U if and only if u^α is subharmonic on U for each $\alpha > 0$. [Hint for the 'if': show that $(u^\alpha - 1)/\alpha$ decreases to $\log u$ as $\alpha \downarrow 0$.]

4. Show that if $\log u$ and $\log v$ are both subharmonic functions on U, then so is $\log(u + v)$.

5. Show that every convex function on $\mathbb{R}$ which is bounded above must be constant. Use this to give another proof of the Liouville theorem (Theorem 2.3.4), that every subharmonic function on $\mathbb{C}$ which is bounded above must be constant.

6. Show that, in the notation of Definition 2.6.7,
$$B_u(r) \geq C_u(r/\sqrt{e}) \quad (0 < r < \rho).$$
[Hint: write $C_u(r)$ as $\psi(\log r)$, where ψ is convex, and apply Jensen's inequality to (2.4).]

7. Show that if $\log u$ is a subharmonic function on the disc $\Delta(0, \rho)$, then $\log M_u(r)$, $\log C_u(r)$ and $\log B_u(r)$ are all convex functions of $\log r$. [Hint: repeat the proof of Theorem 2.6.8 in conjunction with Theorem 2.6.5.]

8. (i) Let $(a_j)_{j\geq 0}$ and $(b_j)_{j\geq 0}$ be non-negative numbers, and for $k \geq 0$ define $c_k = \sum_{j=0}^{k} a_j b_{k-j}$. Use the Cauchy–Schwarz inequality to show that

$$\sum_{k=0}^{\infty} \left(\frac{c_k^2}{k+1}\right) \leq \left(\sum_{j=0}^{\infty} a_j^2\right)\left(\sum_{j=0}^{\infty} b_j^2\right).$$

(ii) Let f and g be holomorphic functions on $\Delta(0,\rho)$. By expanding them as Taylor series, prove that for $0 < r < \rho$,

$$\frac{1}{\pi r^2} \int_{\Delta(0,r)} |f|^2 |g|^2 \, dA$$
$$\leq \left(\frac{1}{2\pi} \int_0^{2\pi} |f(re^{it})|^2 \, dt\right)\left(\frac{1}{2\pi} \int_0^{2\pi} |g(re^{it})|^2 \, dt\right).$$

(iii) Deduce that if u and v are positive functions on $\Delta(0,\rho)$ such that $\log u$ and $\log v$ are subharmonic, then

$$B_{uv}(r) \leq C_u(r) C_v(r) \quad (0 < r < \rho).$$

[Hint: adapt the idea used in the proof of Theorem 2.6.5.]

(iv) Give a geometric interpretation of this last inequality in the case $u = v = |f'|$, where f is a conformal mapping of $\Delta(0,\rho)$ on to a domain D.

2.7 Smoothing

Although subharmonic functions need not be smooth, indeed sometimes far from it, they can nevertheless always be approximated by others which are smooth. A standard way to do this is to use convolutions.

Definition 2.7.1 Let U be an open subset of $\mathbb{C}$, and for $r > 0$ define

$$U_r = \{z \in U : \mathrm{dist}(z, \partial U) > r\}.$$

Let $u: U \to [-\infty, \infty)$ be a locally integrable function, and let $\phi: \mathbb{C} \to \mathbb{R}$ be a continuous function with $\mathrm{supp}\,\phi \subset \Delta(0,r)$. Then their *convolution* is the function $u * \phi: U_r \to \mathbb{R}$ given by

$$u * \phi(z) = \int_{\mathbb{C}} u(z-w)\phi(w) \, dA(w) \quad (z \in U_r).$$

2.7. SMOOTHING

After a change of variable, we also have

$$u * \phi(z) = \int_{\mathbf{C}} u(w)\phi(z-w)\, dA(w) \quad (z \in U_r).$$

This shows that if $\phi \in C^\infty$, then also $u * \phi \in C^\infty$, since we can differentiate under the integral sign arbitrarily many times.

Theorem 2.7.2 (Smoothing Theorem) *Let u be a subharmonic function on a domain D in $\mathbf{C}$, with $u \not\equiv -\infty$. Let $\chi: \mathbf{C} \to \mathbf{R}$ be a function satisfying:*

$$\chi \in C^\infty, \quad \chi \geq 0, \quad \chi(z) = \chi(|z|), \quad \operatorname{supp}\chi \subset \Delta(0,1), \quad \int_{\mathbf{C}} \chi\, dA = 1.$$

For $r > 0$ define

$$\chi_r(z) = \frac{1}{r^2}\chi\left(\frac{z}{r}\right) \quad (z \in \mathbf{C}).$$

Then $u * \chi_r$ is a C^∞ subharmonic function on D_r for each $r > 0$, and $(u * \chi_r) \downarrow u$ on D as $r \downarrow 0$.

An example of a function χ satisfying the hypotheses is given by

$$\chi(z) = \begin{cases} C\exp(-1/(1-4|z|^2)), & \text{if } |z| < 1/2, \\ 0, & \text{if } |z| \geq 1/2, \end{cases}$$

where C is a constant chosen so that $\int \chi\, dA = 1$.

Proof. By Theorem 2.5.1 u is locally integrable, so $u * \chi_r$ makes sense and is C^∞ on D_r. To show it is subharmonic there, apply Theorem 2.4.8 with $(\Omega, \mu) = (\mathbf{C}, \chi_r dA)$ and $v(z,w) = u(z-w)$.

Now fix $\zeta \in D$. For $0 < r < \operatorname{dist}(\zeta, \partial D)$ we have

$$u * \chi_r(\zeta) = \int_0^{2\pi}\int_0^r u(\zeta - se^{it})r^{-2}\chi(s/r)s\, ds\, dt.$$

Making the substitutions $\sigma = s/r$ and $v(z) = u(\zeta - z)$, this becomes

$$u * \chi_r(\zeta) = 2\pi \int_0^1 C_v(r\sigma)\chi(\sigma)\sigma\, d\sigma.$$

By Theorem 2.6.8 $C_v(r\sigma)$ decreases to $v(0)$ as $r \downarrow 0$. Hence by the monotone convergence theorem $u * \chi_r(\zeta)$ decreases to

$$2\pi \int_0^1 v(0)\chi(\sigma)\sigma\, d\sigma = u(\zeta)\int_{\mathbf{C}} \chi\, dA = u(\zeta).$$

Thus $(u * \chi_r) \downarrow u$ on D. $\square$

Corollary 2.7.3 *Let u be a subharmonic function on an open set U in $\mathbb{C}$, and let D be a relatively compact subdomain of U. Then there exist subharmonic functions $(u_n)_{n \geq 1} \in C^\infty(D)$ such that $u_1 \geq u_2 \geq \cdots \geq u$ on D and $\lim_{n \to \infty} u_n = u$.*

Proof. If $u \equiv -\infty$ on D, take $u_n \equiv -n$. Otherwise, choose $r > 0$ such that $D \subset U_r$, and then take $u_n = u * \chi_{r/n}$. $\square$

As an application of this result, we can extend Corollary 2.4.3 to general holomorphic maps.

Theorem 2.7.4 *Let $f: U_1 \to U_2$ be a holomorphic map between open subsets U_1, U_2 of $\mathbb{C}$. If u is subharmonic on U_2, then $u \circ f$ is subharmonic on U_1.*

Proof. Let D_1 be a relatively compact subdomain of U_1. It is enough to show that $u \circ f$ is subharmonic on D_1. Set $D_2 = f(D_1)$ and choose subharmonic functions $(u_n)_{n \geq 1} \in C^\infty(D_2)$ such that $u_n \downarrow u$ on D_2. By Theorem 2.4.4 $\Delta u_n \geq 0$ on D_2 for each n. Now an easy computation gives

$$\Delta(u_n \circ f) = ((\Delta u_n) \circ f)|f'|^2 \quad \text{on } D_1.$$

Hence $\Delta(u_n \circ f) \geq 0$ on D_1, and applying Theorem 2.4.4 the other way we deduce that $u_n \circ f$ is subharmonic there. Letting $n \to \infty$ and using Theorem 2.4.6, it follows that $u \circ f$ is subharmonic on D_1. $\square$

Theorem 2.7.2 can also be used to prove a form of identity principle for subharmonic functions which, although rather weak, is still very useful.

Theorem 2.7.5 (Weak Identity Principle) *Suppose that u and v are subharmonic functions on an open set U in $\mathbb{C}$ such that $u = v$ a.e. on U. Then $u \equiv v$ on U.*

Proof. Suppose first that u and v are bounded below on U. Taking χ as in Theorem 2.7, we then have $u * \chi_r = v * \chi_r$ on U_r, and letting $r \to 0$ we deduce that $u = v$ on U.

The general case follows by applying the one above to $u_n := \max(u, -n)$ and $v_n := \max(v, -n)$, and then letting $n \to \infty$. $\square$

We cannot hope for an identity principle as strong as that for harmonic functions (Theorem 1.1.7): for example, $u(z) := \max(\operatorname{Re} z, 0)$ and $v(z) := 0$ agree on an open subset of $\mathbb{C}$ without being equal on the whole of $\mathbb{C}$. In fact, as we shall see, it is this very lack of rigidity that makes subharmonic functions such a useful tool.

Exercises 2.7

1. Let U be an open subset of $\mathbb{C}$, and let $u\colon U \to [-\infty, \infty)$ be a Borel-measurable function which is locally bounded above and below, and which satisfies the local submean inequality (but is not necessarily upper semicontinuous).

    (i) With χ as in Theorem 2.7.2, show that $u * \chi_r$ is subharmonic on U_r for each $r > 0$, and that $\lim_{r \to 0} u * \chi_r = u^*$, where u^* is the *upper semicontinuous regularization* of u, given by
    $$u^*(z) = \lim_{r \to 0}\left(\sup_{\Delta(z,r)} u\right) \quad (z \in U).$$

    (ii) Show that if $r, s > 0$, then
    $$(u * \chi_r) * \chi_s = (u * \chi_s) * \chi_r \quad \text{on } U_{r+s}.$$
    Deduce that $u * \chi_r$ decreases with r, and that $u * \chi_r = u^* * \chi_r$ on U_r for each $r > 0$.

    (iii) Conclude that u^* is subharmonic on U, and that $u^* = u$ almost everywhere on U.

    (iv) By considering $u_n := \max(u, -n)$ ($n \geq 1$), show that the conclusions in (iii) remain valid if we drop the assumption that u is locally bounded below.

2. Let u be a subharmonic function on an open set U in $\mathbb{C}$, and let v be an upper semicontinuous function on U such that $u \leq v$ almost everywhere. Prove that $u \leq v$ everywhere.

3. Show that the Phragmén–Lindelöf principle (Theorem 2.3.2) remains true even if, instead of assuming v to be finite, we merely suppose that $v \not\equiv +\infty$.

Notes on Chapter 2

§2.3

Exercise 2 is a version of the so-called the Hopf lemma. It has many applications: the one in Exercise 3 is the boundary Schwarz lemma of Burns and Krantz [20]. Exercise 4 is taken from Partington [46], where it is used to estimate the norms of resolvents of operators, and also to prove a Wirtinger-type inequality. The version of the Phragmén–Lindelöf principle outlined in Exercise 5 is due to Carlson. Its application in Exercise 6 is one of a vast range of results relating the growth of holomorphic functions to the distribution of their zeros. For more on this subject see the books of Boas [16] and Levin [41].

§2.6

The book of Radó [53] gives a more extensive treatment of convexity, especially regarding so-called *PL-functions*, namely functions u for which $\log u$ is subharmonic. Among other things, it proves (§3.12) that u is PL if and only if $u(z)|e^{\alpha z}|$ is subharmonic for each $\alpha \in \mathbb{C}$, thereby strengthening Theorem 2.6.5. Several of the exercises are also taken from this source. Exercise 6 is a result of Beardon [11], and Exercise 8 is based on a paper of Carleman [21].

Chapter 3

Potential Theory

3.1 Potentials

Potentials play at least two rôles. Firstly they provide an important source of examples of subharmonic functions, giving us the means, for instance, of constructing such functions with various prescribed properties. Secondly, despite their apparently rather special nature, which makes them comparatively easy to study, we shall see that potentials turn out to be almost as general as arbitrary subharmonic functions, and for many purposes the two classes are equivalent.

We shall define potentials only for finite measures of compact support. (For the definition of the support $\operatorname{supp} \mu$ of a measure μ, see Section A.1 of the Appendix.)

Definition 3.1.1 Let μ be a finite Borel measure on $\mathbb{C}$ with compact support. Its *potential* is the function $p_\mu \colon \mathbb{C} \to [-\infty, \infty)$ defined by

$$p_\mu(z) = \int \log|z - w| \, d\mu(w) \quad (z \in \mathbb{C}).$$

Theorem 3.1.2 *With this notation, p_μ is subharmonic on $\mathbb{C}$, and harmonic on $\mathbb{C} \setminus (\operatorname{supp} \mu)$. Also*

$$p_\mu(z) = \mu(\mathbb{C}) \log|z| + O(|z|^{-1}) \quad \text{as } z \to \infty.$$

Proof. Set $K = \operatorname{supp} \mu$, so μ can be regarded as a measure on K. Applying Theorem 2.4.8 with $v(z, w) = \log|z - w|$ on $\mathbb{C} \times K$, we see that p_μ is subharmonic on $\mathbb{C}$. Applying the same theorem, but with $v(z, w) = -\log|z - w|$ on $(\mathbb{C} \setminus K) \times K$, we also find that p_μ is superharmonic on $\mathbb{C} \setminus K$, and hence harmonic there.

For the last part, observe that for $z \neq 0$,
$$p_\mu(z) = \mu(\mathbf{C})\log|z| + \int \log|1 - w/z|\,d\mu(w).$$
As μ has compact support, the final term is $O(|z|^{-1})$ as $z \to \infty$. □

Potentials enjoy several properties over and above those displayed by general subharmonic functions. We now prove two of these: the continuity principle and the minimum principle.

Theorem 3.1.3 (Continuity Principle) *Let μ be a finite Borel measure on $\mathbf{C}$ with compact support K.*

(a) *If $\zeta_0 \in K$, then $\liminf\limits_{z \to \zeta_0} p_\mu(z) = \liminf\limits_{\substack{\zeta \to \zeta_0 \\ \zeta \in K}} p_\mu(\zeta).$*

(b) *If further $\lim\limits_{\substack{\zeta \to \zeta_0 \\ \zeta \in K}} p_\mu(\zeta) = p_\mu(\zeta_0)$, then $\lim\limits_{z \to \zeta_0} p_\mu(z) = p_\mu(\zeta_0)$.*

Proof. (a) If $p_\mu(\zeta_0) = -\infty$, then by upper semicontinuity $\lim_{z \to \zeta_0} p_\mu(z) = -\infty$ and the result is clear. Thus we can suppose that $p_\mu(\zeta_0) > -\infty$. Then necessarily $\mu(\{\zeta_0\}) = 0$ and so, given $\epsilon > 0$, there exists $r > 0$ such that $\mu(\Delta(\zeta_0, r)) < \epsilon$. Given $z \in \mathbf{C}$, choose $\zeta \in K$ minimizing $|\zeta - z|$. Then for all $w \in K$,
$$\frac{|\zeta - w|}{|z - w|} \leq \frac{|\zeta - z| + |z - w|}{|z - w|} \leq 2.$$

Therefore
$$\begin{aligned} p_\mu(z) &= p_\mu(\zeta) - \int_K \log\left|\frac{\zeta - w}{z - w}\right| d\mu(w) \\ &\geq p_\mu(\zeta) - \epsilon\log 2 - \int_{K \setminus \Delta(\zeta_0, r)} \log\left|\frac{\zeta - w}{z - w}\right| d\mu(w).\end{aligned}$$

As $z \to \zeta_0$ in $\mathbf{C}$, the corresponding $\zeta \to \zeta_0$ in K, and hence
$$\liminf_{z \to \zeta_0} p_\mu(z) \geq \liminf_{\substack{\zeta \to \zeta_0 \\ \zeta \in K}} p_\mu(\zeta) - \epsilon\log 2 - 0.$$

Since ϵ is arbitrary, the result follows.

(b) If p_μ satisfies the premise of (b), then by part (a)
$$\liminf_{z \to \zeta_0} p_\mu(z) = p_\mu(\zeta_0).$$

3.2. POLAR SETS

Also, as p_μ is upper semicontinuous,

$$\limsup_{z \to \zeta_0} p_\mu(z) \leq p_\mu(\zeta_0).$$

Combining these observations, we deduce that p_μ satisfies the conclusion of (b). □

Theorem 3.1.4 (Minimum Principle) *Let μ be a finite Borel measure on $\mathbb{C}$ with compact support K. If $p_\mu \geq M$ on K, then $p_\mu \geq M$ on the whole of $\mathbb{C}$.*

Proof. Put $u = -p_\mu$ on $\mathbb{C} \setminus K$. Then u is subharmonic on $\mathbb{C} \setminus K$ and (assuming that $\mu \neq 0$) $u(z) \to -\infty$ as $z \to \infty$. Also if $\zeta_0 \in \partial K$, then by Theorem 3.1.3 (a)

$$\limsup_{\substack{z \to \zeta_0 \\ z \in \mathbb{C} \setminus K}} u(z) \leq -\liminf_{z \to \zeta_0} p_\mu(z) = -\liminf_{\substack{\zeta \to \zeta_0 \\ \zeta \in K}} p_\mu(\zeta) \leq -M.$$

Applying the maximum principle to u on each component of $\mathbb{C} \setminus K$, we get $u \leq -M$ there. Hence $p_\mu \geq M$ on $\mathbb{C}$. □

Exercises 3.1

1. Give an example of a finite Borel measure μ on $\mathbb{C}$ with compact support K such that, for some $\zeta_0 \in K$,

$$\limsup_{z \to \zeta_0} p_\mu(z) > \limsup_{\substack{\zeta \to \zeta_0 \\ \zeta \in K}} p_\mu(\zeta).$$

[Hint: look at the function in Exercise 2.2.2.]

3.2 Polar Sets

Polar sets play the rôle of negligible sets in potential theory, much as sets of measure zero do in measure theory. To define them, we first need to introduce the notion of energy.

Definition 3.2.1 Let μ be a finite Borel measure on $\mathbb{C}$ with compact support. Its *energy* $I(\mu)$ is given by

$$I(\mu) := \iint \log|z - w|\, d\mu(z)\, d\mu(w) = \int p_\mu(z)\, d\mu(z).$$

To explain this terminology, think of μ as being a charge distribution on $\mathbb{C}$. Then $p_\mu(z)$ represents the potential energy at z due to μ, and so the total energy of μ is just $\int p_\mu(z)\, d\mu(z)$, in other words $I(\mu)$. (Actually, since like charges repel, most physicists would define the energy as $-I(\mu)$, but Definition 3.2.1 will be more convenient for us.)

It is possible that $I(\mu) = -\infty$. Indeed some sets only support measures of infinite energy. These are important enough to deserve a name.

Definition 3.2.2 (a) A subset E of $\mathbb{C}$ is called *polar* if $I(\mu) = -\infty$ for every finite Borel measure $\mu \neq 0$ for which $\operatorname{supp}\mu$ is a compact subset of E.

(b) A property is said to hold *nearly everywhere (n.e.)* on a subset S of $\mathbb{C}$ if it holds everywhere on $S \setminus E$, for some Borel polar set E.

Clearly singleton sets are polar. Also every subset of a polar set is polar. In the other direction, if a set is non-polar, then it contains a compact subset which is non-polar (namely $\operatorname{supp}\mu$, for some measure μ with $I(\mu) > -\infty$).

It is easy to see that measures of finite energy can have no atoms. In fact, more generally, they do not charge polar sets.

Theorem 3.2.3 *Let μ be a finite Borel measure on $\mathbb{C}$ with compact support, and suppose that $I(\mu) > -\infty$. Then $\mu(E) = 0$ for every Borel polar set E.*

Proof. Let E be a Borel set such that $\mu(E) > 0$. We shall show that E is not polar. By regularity of μ, we can choose a compact subset K of E with $\mu(K) > 0$ (see Section A.2 of the Appendix). Set $\widetilde{\mu} = \mu|_K$ and $d = \operatorname{diam}(\operatorname{supp}\mu)$. Then $\widetilde{\mu}$ is a finite non-zero measure whose support is a compact subset of E, and

$$\begin{aligned}
I(\widetilde{\mu}) &= \int_K \int_K \log\left|\frac{z-w}{d}\right| d\mu(z)\, d\mu(w) + \mu(K)^2 \log d \\
&\geq \int_\mathbb{C} \int_\mathbb{C} \log\left|\frac{z-w}{d}\right| d\mu(z)\, d\mu(w) + \mu(K)^2 \log d \\
&= I(\mu) - \mu(\mathbb{C})^2 \log d + \mu(K)^2 \log d > -\infty.
\end{aligned}$$

Hence E is non-polar, as claimed. $\square$

Corollary 3.2.4 *Every Borel polar set has Lebesgue measure zero.*

Proof. It is enough to show that, for $\rho > 0$, the measure $d\mu := dA|_{\Delta(0,\rho)}$ has energy $I(\mu) > -\infty$. For then by Theorem 3.2.3, every Borel polar set E

3.2. POLAR SETS

has μ-measure zero, i.e. $E \cap \Delta(0,\rho)$ has Lebesgue measure zero, and the result then follows by letting $\rho \to \infty$.

Accordingly, fix $\rho > 0$ and let $d\mu = dA|_{\Delta(0,\rho)}$. Then for $z \in \Delta(0,\rho)$,

$$\begin{aligned}p_\mu(z) &= \int_{\Delta(0,\rho)} \log\left|\frac{z-w}{2\rho}\right| dA(w) + \pi\rho^2 \log(2\rho) \\ &\geq \int_{t=0}^{2\pi}\int_{r=0}^{2\rho} \log\left(\frac{r}{2\rho}\right) r\, dr\, dt + \pi\rho^2 \log(2\rho) \\ &= -2\pi\rho^2 + \pi\rho^2 \log(2\rho).\end{aligned}$$

Hence we have

$$I(\mu) = \int_{\Delta(0,\rho)} p_\mu(z)\, d\mu(z) \geq (-2\pi\rho^2 + \pi\rho^2 \log(2\rho))\pi\rho^2 > -\infty,$$

as desired. □

Thus 'nearly everywhere' implies 'almost everywhere'. Exercise 1 below shows that the converse is false. In fact an argument similar to the one above, but rather more technical, shows that Borel polar sets actually have α-dimensional Hausdorff measure zero for each $\alpha > 0$, and thus are of Hausdorff dimension zero. We shall not give the details here.

Corollary 3.2.5 *A countable union of Borel polar sets is polar. In particular, every countable subset of $\mathbb{C}$ is polar.*

This result may fail if the sets are not Borel. An example is outlined in Exercise 2.

Proof. Suppose that $(E_n)_{n\geq 1}$ are Borel polar sets and that $E = \cup_n E_n$. Let μ be a finite Borel measure on $\mathbb{C}$ whose support is a compact subset of E. If $I(\mu) > -\infty$, then by Theorem 3.2.3 $\mu(E_n) = 0$ for each n, so $\mu(E) = 0$, and hence $\mu = 0$. This shows that E is polar. □

We conclude by remarking that, though every countable set is polar, not every polar set is countable. This will be demonstrated in Section 3.5, and concrete examples of uncountable polar sets will follow in Section 5.3.

Exercises 3.2

1. Adapt the proof of Corollary 3.2.4 to show that every Borel polar subset of $\mathbb{R}$ has one-dimensional Lebesgue measure zero.

58   CHAPTER 3. POTENTIAL THEORY

2. (This question requires some background in set theory.)

   (i) Let S be a set, and let $\mathcal{T}$ be a collection of infinite subsets of S such that $\text{card}(T) \geq \text{card}(\mathcal{T})$ for all $T \in \mathcal{T}$. Show that S can be partitioned into subsets P and Q neither of which contains any element of $\mathcal{T}$. [Hint: Let β be the smallest ordinal with $\text{card}(\beta) = \text{card}(\mathcal{T})$, and index $\mathcal{T}$ as $(T_\alpha)_{\alpha < \beta}$. Construct P and Q inductively.]

   (ii) Let $\mathcal{T}$ be the collection of all uncountable compact subsets of $\mathbb{C}$. Show that $\text{card}(\mathcal{T}) = \mathbf{c}$, and that $\text{card}(T) = \mathbf{c}$ for all $T \in \mathcal{T}$.

   (iii) Deduce that $\mathbb{C}$ can be partitioned into subsets P and Q such that each compact subset of P or Q is countable.

   (iv) Conclude that the union of two non-Borel polar sets need not be polar.

3.3 Equilibrium Measures

In physics, a charge placed upon a conductor will distribute itself so as to minimize the energy. In our context, this suggests looking at probability measures μ on a compact set K which maximize $I(\mu)$. Not only are they of physical relevance, but they turn out to be mathematically very useful too.

Definition 3.3.1 Let K be a compact subset of $\mathbb{C}$, and denote by $\mathcal{P}(K)$ the collection of all Borel probability measures on K. If there exists $\nu \in \mathcal{P}(K)$ such that
$$I(\nu) = \sup_{\mu \in \mathcal{P}(K)} I(\mu),$$
then ν is called an *equilibrium measure* for K.

Theorem 3.3.2 *Every compact set K in $\mathbb{C}$ has an equilibrium measure.*

We shall see later (Theorem 3.7.6) that in fact this equilibrium measure is unique, provided that K is non-polar. (Of course if K is polar then every $\mu \in \mathcal{P}(K)$ is an equilibrium measure since they all satisfy $I(\mu) = -\infty$.)

To prove Theorem 3.3.2, we shall need the notion of weak*-convergence of probability measures. This is defined in Section A.4 of the Appendix, where it is also shown that every sequence (μ_n) in $\mathcal{P}(K)$ has a weak*-convergent subsequence.

Lemma 3.3.3 *If $\mu_n \xrightarrow{w^*} \mu$ in $\mathcal{P}(K)$, then $\limsup_{n \to \infty} I(\mu_n) \leq I(\mu)$.*

3.3. EQUILIBRIUM MEASURES

Proof. Given continuous functions ϕ, ψ on K, the definition of weak*-convergence implies that, as $n \to \infty$,

$$\int_K \int_K \phi(z)\psi(w)\, d\mu_n(z)\, d\mu_n(w) \to \int_K \int_K \phi(z)\psi(w)\, d\mu(z)\, d\mu(w).$$

Now using the Stone–Weierstrass theorem, one can show that every continuous function $\chi(z,w)$ on $K \times K$ can be uniformly approximated by finite sums of the form $\sum_j \phi_j(z)\psi_j(w)$, where the ϕ_j, ψ_j are continuous functions on K. It follows that for every such χ,

$$\int_K \int_K \chi(z,w)\, d\mu_n(z)\, d\mu_n(w) \to \int_K \int_K \chi(z,w)\, d\mu(z)\, d\mu(w) \quad \text{as } n \to \infty.$$

Applying this with $\chi(z,w) := \max(\log|z-w|, -m)$, where $m \geq 1$, we get

$$\begin{aligned}
\limsup_{n\to\infty} I(\mu_n) &= \limsup_{n\to\infty} \int_K \int_K \log|z-w|\, d\mu_n(z)\, d\mu_n(w) \\
&\leq \limsup_{n\to\infty} \int_K \int_K \max(\log|z-w|, -m)\, d\mu_n(z)\, d\mu_n(w) \\
&= \int_K \int_K \max(\log|z-w|, -m)\, d\mu(z)\, d\mu(w).
\end{aligned}$$

The result follows upon letting $m \to \infty$ and using the monotone convergence theorem. $\square$

Proof of Theorem 3.3.2. Put $M = \sup_{\mu \in \mathcal{P}(K)} I(\mu)$, and choose a sequence (μ_n) in $\mathcal{P}(K)$ such that $I(\mu_n) \to M$ as $n \to \infty$. By Theorem A.4.2 in the Appendix, there is a subsequence (μ_{n_k}) which is weak*-convergent to some $\nu \in \mathcal{P}(K)$. By Lemma 3.3.3

$$I(\nu) \geq \limsup_{k \to \infty} I(\mu_{n_k}) = M,$$

so ν is an equilibrium measure for K. $\square$

Physical intuition would tend to suggest that if ν is an equilibrium measure for K then p_ν should be constant on K (for otherwise charge would flow from one part of K to another, disturbing the equilibrium). This idea is confirmed by the next theorem, and even serves to motivate the proof.

Theorem 3.3.4 (Frostman's Theorem) *Let K be a compact set in $\mathbb{C}$, and let ν be an equilibrium measure for K. Then:*

(a) $p_\nu \geq I(\nu)$ *on $\mathbb{C}$;*

(b) $p_\nu = I(\nu)$ *on $K \setminus E$, where E is an F_σ polar subset of ∂K.*

It can happen that the exceptional set E is non-empty. An example is outlined in Exercise 1 below.

Proof. If $I(\nu) = -\infty$ (i.e. K is polar) then the result is obvious, so we may as well assume that $I(\nu) > -\infty$. It is sufficient to prove that

(i) $K_n := \{z \in K : p_\nu(z) \geq I(\nu) + 1/n\}$ is polar for each $n \geq 1$, and

(ii) $L_n := \{z \in \operatorname{supp}\nu : p_\nu(z) < I(\nu) - 1/n\}$ is empty for each $n \geq 1$.

For (ii) then implies that $p_\nu \geq I(\nu)$ on $\operatorname{supp}\nu$, and so by the minimum principle (Theorem 3.1.4) we get $p_\nu \geq I(\nu)$ on $\mathbb{C}$, which gives (a). Also, if we put $E = \cup_n K_n$, then (i) together with Corollary 3.2.5 implies that E is an F_σ polar set. Since $p_\nu \leq I(\nu)$ on $K \setminus E$, this gives (b), apart from the assertion that $E \subset \partial K$. This last is proved by observing that as E is polar, it must have Lebesgue measure zero, so $p_\nu = I(\nu)$ a.e. on K, and hence by the weak identity principle (Theorem 2.7.5) $p_\nu = I(\nu)$ everywhere on $\operatorname{int}(K)$.

It thus remains to prove (i) and (ii). We shall prove (i) by contradiction. Suppose, if possible, that some K_n is non-polar. Choose $\mu \in \mathcal{P}(K_n)$ with $I(\mu) > -\infty$. Since $I(\nu) = \int p_\nu \, d\nu$, there exists $z_0 \in \operatorname{supp}\nu$ such that $p_\nu(z_0) \leq I(\nu)$. By upper semicontinuity, there exists $r > 0$ such that $p_\nu < I(\nu) + 1/2n$ on $\overline{\Delta}(z_0, r)$. In particular, $\overline{\Delta}(z_0, r) \cap K_n = \emptyset$. As $z_0 \in \operatorname{supp}\nu$, the number $a := \nu(\overline{\Delta}(z_0, r))$ is strictly positive. Define a signed measure σ on K by

$$\sigma = \begin{cases} \mu & \text{on } K_n, \\ -\nu/a & \text{on } \overline{\Delta}(z_0, r), \\ 0 & \text{otherwise.} \end{cases}$$

Then for each $t \in (0, a)$, the measure $\nu_t := \nu + t\sigma$ is positive, and therefore $\nu_t \in \mathcal{P}(K)$. Also, noting that $I(\mu) > -\infty$ implies $I(|\sigma|) > -\infty$, we have

$$\begin{aligned} I(\nu_t) &- I(\nu) \\ &= 2t \iint \log|z - w| \, d\nu(w) \, d\sigma(z) + t^2 \iint \log|z - w| \, d\sigma(w) \, d\sigma(z) \\ &= 2t \int p_\nu(z) \, d\sigma(z) + O(t^2) \\ &= 2t \left(\int_{K_n} p_\nu(z) \, d\mu(z) - \int_{\overline{\Delta}(z_0, r)} p_\nu(z) \, d\nu(z)/a + O(t) \right) \\ &\geq 2t \left(\left(I(\nu) + \frac{1}{n} \right) - \left(I(\nu) + \frac{1}{2n} \right) + O(t) \right). \end{aligned}$$

Therefore $I(\nu_t) > I(\nu)$ if t is sufficiently small, contradicting the assumption that ν is an equilibrium measure. Hence each K_n is polar, which proves (i).

3.3. EQUILIBRIUM MEASURES

We shall also prove (ii) by contradiction. Suppose, if possible, that some L_n is non-empty. Pick $z_1 \in L_n$. By upper semicontinuity, there exists $s > 0$ such that $p_\nu < I(\nu) - 1/n$ on $\overline{\Delta}(z_1, s)$. As $z_1 \in \operatorname{supp} \nu$, the number $b := \nu(\overline{\Delta}(z_1, s))$ is strictly positive. Now by (i) and Corollary 3.2.4, $\nu(K_n) = 0$ for each n, and so $p_\nu \leq I(\nu)$ ν-almost everywhere on K. Hence

$$\begin{aligned}
I(\nu) &= \int_K p_\nu \, d\nu \\
&= \int_{\overline{\Delta}(z_1,s)} p_\nu \, d\nu + \int_{K \setminus \overline{\Delta}(z_1,s)} p_\nu \, d\nu \\
&\leq \left(I(\nu) - \frac{1}{n} \right) b + I(\nu)(1 - b) \\
&< I(\nu),
\end{aligned}$$

which is obviously a contradiction. Hence each L_n is empty, which gives (ii) and completes the proof. $\square$

Frostman's theorem is very important, serving many different purposes. Indeed, it is sometimes referred to as the 'fundamental theorem of potential theory'—a grandiose title but, as we shall see, one that is fully justified.

Exercises 3.3

1. Let K be a compact set of the form $\overline{\Delta} \cup E$, where $\overline{\Delta}$ is a closed disc, and E is a polar subset of $\mathbb{C} \setminus \overline{\Delta}$. Let ν be an equilibrium measure for K.

    (i) Explain why $I(\nu) > -\infty$.

    (ii) Show that $\nu(E) = 0$, and deduce that p_ν is harmonic on $\mathbb{C} \setminus \overline{\Delta}$.

    (iii) Use the maximum principle to show that $p_\nu > I(\nu)$ on E.

2. Let μ be a Borel probability measure on $\mathbb{C}$ with compact support, let K be a compact subset of $\mathbb{C}$, and let ν be an equilibrium measure for K.

    (i) Show that $\sup_K p_\mu \geq I(\nu)$.

    (ii) Show further that if $\operatorname{supp} \mu \subset K$ then $\inf_K p_\mu \leq I(\nu)$.

    [Hint: by Fubini's theorem $\int p_\mu \, d\nu = \int p_\nu \, d\mu$.]

3.4 Upper Semicontinuous Regularization

We saw in Theorem 2.4.6 that the limit of a decreasing sequence of subharmonic functions is subharmonic. At the same time, we remarked that the corresponding result for an increasing sequence was false, because the limit might not be upper semicontinuous. One way round this problem is to make the limit upper semicontinuous by regularizing it.

Definition 3.4.1 Let X be a topological space, and let $u\colon X \to [-\infty, \infty)$ be a function which is locally bounded above on X. Its *upper semicontinuous regularization* $u^*\colon X \to [-\infty, \infty)$ is defined by

$$u^*(x) := \limsup_{y \to x} u(y) = \inf_N \Bigl(\sup_{y \in N} u(y)\Bigr) \quad (x \in X),$$

the infimum being taken over all neighbourhoods N of x.

It is easily checked that u^* is an upper semicontinuous function on X such that $u^* \geq u$, and also that it is the least such function.

Returning to our problem about an increasing sequence of subharmonic functions, it is perhaps not too surprising to learn that provided the limit u is locally bounded above, its upper semicontinuous regularization u^* is subharmonic. What is much less obvious is that u^* is very nearly equal to u. From Exercise 2.7.1 it follows that the two functions must agree almost everywhere, and in fact much more than this is true.

Theorem 3.4.2 (Brelot–Cartan Theorem) *Let $\mathcal{V}$ be a collection of subharmonic functions on an open subset U of $\mathbb{C}$, and suppose that the function $u := \sup_{v \in \mathcal{V}} v$ is locally bounded above on U. Then:*

(a) *u^* is subharmonic on U;*

(b) *$u^* = u$ nearly everywhere on U.*

Part (b) says that $u^* = u$ everywhere on U outside some Borel polar set. Note however that the set $\{z : u^*(z) \neq u(z)\}$ itself might not be Borel, since $\mathcal{V}$ could be uncountable.

Proof. (a) Suppose that $\overline{\Delta}(w, \rho) \subset U$. Then for each $v \in \mathcal{V}$,

$$v(w) \leq \frac{1}{2\pi} \int_0^{2\pi} v(w + \rho e^{i\theta})\, d\theta \leq \frac{1}{2\pi} \int_0^{2\pi} u^*(w + \rho e^{i\theta})\, d\theta.$$

Taking the supremum over all $v \in \mathcal{V}$, we get

$$(3.1) \qquad u(w) \leq \frac{1}{2\pi} \int_0^{2\pi} u^*(w + \rho e^{i\theta})\, d\theta.$$

3.4. UPPER SEMICONTINUOUS REGULARIZATION

Now choose $w_n \to w$ such that $\lim_{n\to\infty} u(w_n) = u^*(w)$. If n is sufficiently large, then $\overline{\Delta}(w_n, \rho) \subset U$, so (3.1) holds with w replaced by w_n throughout. Using Fatou's lemma, it follows that

$$u^*(w) \leq \frac{1}{2\pi}\int_0^{2\pi} \limsup_{n\to\infty} u^*(w_n + \rho e^{i\theta})\, d\theta \leq \frac{1}{2\pi}\int_0^{2\pi} u^*(w + \rho e^{i\theta})\, d\theta.$$

Thus u^* satisfies the submean inequality, and so is subharmonic on U.

(b) We first consider the case when $\mathcal{V}$ is countable, so that u is Borel-measurable. Now the set $\{z \in U : u^*(z) \neq u(z)\}$ can be written as a countable union of Borel sets of the form

$$E := \{z \in \Delta : u(z) \leq \beta < u^*(z)\},$$

where Δ is a disc such that $\overline{\Delta} \subset U$, and β is a rational number. Thus it suffices to show that each such set E is polar. We shall do this by contradiction. Suppose, if possible, that for some Δ, β the set E is non-polar. Then E contains a compact non-polar subset K, say. Let ν be an equilibrium measure for K, and define $q\colon \mathbb{C} \to [-\infty, \infty)$ by

$$q = C(p_\nu - I(\nu)) + \beta,$$

where C is a positive constant chosen large enough so that $\inf_{\partial \Delta} q > \sup_{\partial \Delta} u$. (Such a choice is possible, since by Frostman's theorem and the maximum principle $p_\nu > I(\nu)$ on the unbounded component of $\mathbb{C} \setminus K$.) Then for each $v \in \mathcal{V}$, the function $v - q$ is subharmonic on $\Delta \setminus K$, and if $\zeta \in \partial(\Delta \setminus K)$ then

$$\limsup_{z \to \zeta}(v - q)(z) \leq \left\{\begin{array}{ll} u(\zeta) - \inf_{\partial\Delta} q, & \text{if } \zeta \in \partial\Delta, \\ u(\zeta) - \beta, & \text{if } \zeta \in \partial K, \end{array}\right\} \leq 0.$$

Hence by the maximum principle $v \leq q$ on $\Delta \setminus K$. Therefore $u \leq q$ on $\Delta \setminus K$. Also $u \leq \beta \leq q$ on K, so in fact $u \leq q$ on the whole of Δ, and hence $u^* \leq q$ on Δ. This implies that $q > \beta$ on K, or in other words $p_\nu > I(\nu)$ on K, which contradicts Theorem 3.3.4 (b). Thus E is polar, as required.

We now turn to the case when $\mathcal{V}$ is uncountable. Choose a countable base (D_j) of relatively compact open subsets of U. For each pair $j, k \geq 1$, there exists $v_{jk} \in \mathcal{V}$ such that

$$\sup_{D_j} v_{jk} > \sup_{D_j} u^* - 1/k.$$

If we set $u_0 = \sup_{j,k} v_{jk}$, then $u_0 \leq u$ and $u_0^* = u^*$. By the countable case $u_0^* = u_0$ n.e. on U. Hence it follows that $u^* = u$ n.e. on U. $\square$

Of course, the Brelot–Cartan theorem applies in particular to limits of increasing sequences. There is also a corresponding result for more general sequences.

Theorem 3.4.3 *Let $(u_n)_{n\geq 1}$ be a sequence of subharmonic functions on an open set U, and suppose that $\sup_n u_n$ is locally bounded above on U. If $u = \limsup_{n\to\infty} u_n$, then:*

(a) *u^* is subharmonic on U;*

(b) *$u^* = u$ nearly everywhere on U;*

(c) *if $\phi: U \to \mathbf{R}$ is continuous and $\phi \geq u$, then $\max(u_n, \phi) \to \phi$ locally uniformly on U as $n \to \infty$.*

Proof. (a) If $\overline{\Delta}(w, \rho) \subset U$, then for each $n \geq 1$

$$u_n(w) \leq \frac{1}{2\pi} \int_0^{2\pi} u_n(w + \rho e^{i\theta})\, d\theta.$$

Taking $\limsup_{n\to\infty}$ of both sides and using Fatou's lemma, we get

$$u(w) \leq \frac{1}{2\pi} \int_0^{2\pi} u(w + \rho e^{i\theta})\, d\theta \leq \frac{1}{2\pi} \int_0^{2\pi} u^*(w + \rho e^{i\theta})\, d\theta.$$

The same argument as used in Theorem 3.4.2 (a) now shows that u^* is subharmonic on U.

(b) For each $n \geq 1$ put $v_n = \sup_{m\geq n} u_m$. Then $v_n \downarrow u$, and $v_n^* \downarrow v$ say, where $v \geq u^* \geq u$. Now by Theorem 3.4.2 (b) $v_n^* = v_n$ n.e. for each n, therefore $v = u$ n.e., and hence $u^* = u$ n.e.

(c) Since $\phi \leq \max(u_n, \phi) \leq \max(v_n^*, \phi)$ for each n, it suffices to prove that $\max(v_n^*, \phi) \to \phi$ uniformly on compact sets. As (v_n^*) is a decreasing sequence of upper semicontinuous functions, by Dini's theorem this will be true provided that $\lim_{n\to\infty} v_n^* \leq \phi$, which we shall now prove.

By Theorem 3.4.2 (a) each v_n^* is subharmonic on U, and since $v_n^* \downarrow v$, it follows from Theorem 2.4.6 that v is subharmonic. Also, from (a) u^* is subharmonic on U, and from (b) $v = u^*$ n.e. (and thus a.e.) on U. Hence by the weak identity principle (Theorem 2.7.5) $v = u^*$ everywhere on U, and so

$$\lim_{n\to\infty} v_n^* = v = u^* \leq \phi^* = \phi,$$

as required. $\square$

Exercises 3.4

1. Use the criterion of Theorem 2.4.1 (c) to give an alternative proof of the first part of the Brelot–Cartan theorem.

2. Let $v: U \to [-\infty, \infty)$ be an upper semicontinuous function on an open set U in $\mathbb{C}$. Show that there is a greatest subharmonic function u on U such that $u \leq v$.

3.5 Minus-Infinity Sets

Earlier we saw (Corollary 2.5.3) that if u is subharmonic on a domain and $u \not\equiv -\infty$, then the set where $u = -\infty$ has Lebesgue measure zero. We are now in a position to prove a much stronger result.

Theorem 3.5.1 *Let u be a subharmonic function on a domain D in $\mathbb{C}$, with $u \not\equiv -\infty$. Then $E := \{z \in D : u(z) = -\infty\}$ is a G_δ polar set.*

Proof. Since $E = \bigcap_n \{z : u(z) < -n\}$, it is certainly a G_δ set. To show it is polar, put $v = \lim_{n \to \infty}(u/n)$, so that

$$v(z) = \begin{cases} 0, & z \in D \setminus E, \\ -\infty, & z \in E. \end{cases}$$

Now by Theorem 3.4.3 (a) v^* is subharmonic on D, and since it evidently attains a maximum value 0 there, it follows that $v^* \equiv 0$ on D. Also by Theorem 3.4.3 (b) $v^* = v$ n.e. on D. Therefore $v = 0$ n.e. on D, and E is indeed polar. □

This result allows us to demonstrate the existence of uncountable polar sets. For example, the set E occurring in Theorem 2.5.4 is uncountable, and by Theorem 3.5.1 it is polar. More concrete examples will appear in Section 5.3.

Theorem 3.5.1 is sharp in the sense that every G_δ polar set arises as the set where some subharmonic function $u = -\infty$. This converse, Deny's theorem, is too hard for us to prove here; instead we content ourselves with the following result which, though weaker, is good enough for most purposes.

Theorem 3.5.2 *Let E be an F_σ polar set, and let F be an F_σ set disjoint from E. Then there exists a subharmonic function $u: \mathbb{C} \to [-\infty, \infty)$ such that $u = -\infty$ on E and $u > -\infty$ on F.*

We shall prove this via a lemma which is of interest in its own right.

Lemma 3.5.3 *Let E be a compact polar set, and let F be a compact set disjoint from E. Then there exists a Borel probability measure μ on $\mathbb{C}$ with compact support such that $E = \{z \in \mathbb{C} : p_\mu(z) = -\infty\}$ and $\operatorname{supp}\mu \cap F = \emptyset$.*

Proof. Let $(K_n)_{n \geq 1}$ be a sequence of compact sets, with $K_{n+1} \subset \operatorname{int}(K_n)$ for all n, such that $\cap_n K_n = E$ and $K_1 \cap F = \emptyset$. For each n, let ν_n be an equilibrium measure for K_n. Note that $I(\nu_n) > -\infty$ since $\operatorname{int}(K_n) \neq \emptyset$. Now $\nu_n \in \mathcal{P}(K_1)$ for all n, so by Theorem A.4.2 of the Appendix there is a subsequence of the (ν_n) (which, by relabelling, we may assume to be the whole sequence) that is weak*-convergent to some $\nu \in \mathcal{P}(K_1)$. In fact, since $\operatorname{supp}\nu_n \subset K_n$ for each n, we must have $\operatorname{supp}\nu \subset E$. As E is polar, it follows that $I(\nu) = -\infty$. Hence by Lemma 3.3.3 $I(\nu_n) \to -\infty$ as $n \to \infty$ and so, replacing (ν_n) by a further subsequence, we can suppose that $I(\nu_n) < -2^n$ for each n. Put $\mu = \sum_1^\infty 2^{-n}\nu_n$. Then $\mu \in \mathcal{P}(K_1)$, so $\operatorname{supp}\mu \cap F = \emptyset$, and to finish the proof we shall show that $p_\mu(z) = -\infty$ if and only if $z \in E$.

First suppose that $z \in E$. Then $z \in \operatorname{int}(K_n)$ for each n, so by Theorem 3.3.4 (b) $p_{\nu_n}(z) = I(\nu_n) < -2^n$. Hence

$$p_\mu(z) = \sum_1^\infty 2^{-n} p_{\nu_n}(z) \leq \sum_1^\infty 2^{-n}(-2^n) = -\infty.$$

Now suppose that $z \notin E$. Choose n_0 such that $z \notin K_{n_0}$ and put $\delta = \operatorname{dist}(z, K_{n_0})$. Then for all $n \geq n_0$

$$p_{\nu_n}(z) \geq \int \log \delta \, d\nu_n = \log \delta,$$

and also by Theorem 3.3.4 (a) $p_{\nu_n}(z) \geq I(\nu_n) > -\infty$ for every n. Hence

$$p_\mu(z) = \sum_1^\infty 2^{-n} p_{\nu_n}(z) \geq \sum_1^{n_0-1} 2^{-n} I(\nu_n) + \sum_{n_0}^\infty 2^{-n} \log \delta > -\infty.$$

This completes the proof of the lemma. □

We remark in passing that it is not clear from the proof above whether μ can be chosen so that $\operatorname{supp}\mu \subset E$. The fact that it can (Evans' theorem) will be proved in Section 5.5.

Proof of Theorem 3.5.2. Write E as $\cup_n E_n$ and F as $\cup_n F_n$, where (E_n) and (F_n) are increasing sequences of compact sets. By Lemma 3.5.3, for each n there exists a Borel probability measure μ_n with compact support such that $E_n = \{z : p_{\mu_n}(z) = -\infty\}$ and $\operatorname{supp}\mu_n \cap F_n = \emptyset$. Then p_{μ_n} is

3.6. REMOVABLE SINGULARITIES

bounded above on $\Delta(0,n)$ and below on F_n, so we can choose constants $\alpha_n > 0$ and $\beta_n \in \mathbb{R}$ such that $u_n := \alpha_n p_{\mu_n} + \beta_n$ satisfies

$$\sup_{\Delta(0,n)} u_n < 0 \quad \text{and} \quad \inf_{F_n} u_n > -2^{-n}.$$

Put $u = \sum_1^\infty u_n$. Then on any bounded set, the sequence of partial sums is eventually decreasing, and so by Theorem 2.4.6 u is subharmonic on $\mathbb{C}$. Also if $z \in E$, then $u_n(z) = -\infty$ for some n, and so $u(z) = -\infty$. Finally if $z \in F$, then $u_n(z) > -\infty$ for all n and $u_n(z) \geq -2^{-n}$ for all sufficiently large n, whence $u(z) > -\infty$. □

We conclude by recording an important special case of Theorem 3.5.2.

Corollary 3.5.4 *If E is a closed polar subset of $\mathbb{C}$, then there exists a subharmonic function u on $\mathbb{C}$ such that $E = \{z \in \mathbb{C} : u(z) = -\infty\}$.*

Proof. Apply Theorem 3.5.2 with $F = \mathbb{C} \setminus E$. □

Exercises 3.5

1. Use the result of Exercise 3.3.2 to give an alternative proof of Theorem 3.5.1 in the case when u is a potential.

3.6 Removable Singularities

In each of the last three sections we have encountered theorems asserting that certain exceptional sets are polar. It is thus of interest to determine in what ways polar sets are 'negligible'. The key to this is the following removable singularity theorem.

Theorem 3.6.1 (Removable Singularity Theorem) *Let U be an open subset of $\mathbb{C}$, let E be a closed polar set, and let u be a subharmonic function on $U \setminus E$. Suppose that each point of $U \cap E$ has a neighbourhood N such that u is bounded above on $N \setminus E$. Then u has a unique subharmonic extension to the whole of U.*

Proof. Uniqueness follows immediately from the weak identity principle (Theorem 2.7.5), since E has measure zero. To construct the extension, we define u on $U \cap E$ by

$$u(w) = \limsup_{\substack{z \to w \\ z \in U \setminus E}} u(z) \quad (w \in U \cap E).$$

The boundedness condition ensures that $u < \infty$ everywhere, and so u is upper semicontinuous on U. To check that it is subharmonic, we shall use Theorem 2.4.1 (c). Let D be a relatively compact subdomain of U, and let h be a harmonic function on D such that $\limsup_{z \to \zeta}(u-h)(z) \leq 0$ for all $\zeta \in \partial D$. We need to show that $u \leq h$ on D. Now by Corollary 3.5.4 there exists a subharmonic function v on $\mathbb{C}$ such that $E = \{z : v(z) = -\infty\}$. For each $\epsilon > 0$, the function $u - h + \epsilon v$ is certainly subharmonic on $D \setminus E$, and equals $-\infty$ on E, so in fact it is subharmonic on the whole of D. Therefore by the maximum principle

$$u - h + \epsilon v \leq \sup_{\partial D}(\epsilon v) \quad \text{on } D.$$

Letting $\epsilon \to 0$ we deduce that $u \leq h$ on $D \setminus E$. From the way that u is defined on $D \cap E$, it follows that $u \leq h$ there too. Hence $u \leq h$ on D, as required. □

Corollary 3.6.2 *Let U be an open subset of $\mathbb{C}$, let E be a closed polar set, and let h be a harmonic function on $U \setminus E$. Suppose that each point of $U \cap E$ has a neighbourhood N such that h is bounded on $N \setminus E$. Then h has a unique harmonic extension to the whole of U.*

Proof. Uniqueness is clear. For the existence, apply Theorem 3.6.1 to $\pm h$ to obtain functions u and v which are subharmonic on U, and which agree respectively with h and $-h$ on $U \setminus E$. Then $u + v$ is subharmonic on U and $u + v = 0$ on $U \setminus E$, so by the weak identity principle $u + v = 0$ on the whole of U. Therefore u is superharmonic on U as well as being subharmonic, and hence it is harmonic there. Thus it is the desired extension of h. □

The removable singularity theorem can be used to demonstrate a further sense in which polar sets are small.

Theorem 3.6.3 *Let D be a domain in $\mathbb{C}$ and let E be a closed polar set. Then $D \setminus E$ is still connected.*

Proof. Suppose that $D \setminus E = A \cup B$, where A and B are disjoint open sets. Define $u: D \setminus E \to [-\infty, \infty)$ by

$$u = \begin{cases} 0 & \text{on } A, \\ -\infty & \text{on } B. \end{cases}$$

By Theorem 3.6.1 u has a subharmonic extension to the whole of D. It then follows from Corollary 2.5.3 that if $B \neq \emptyset$ then $u \equiv -\infty$ on D, and so $A = \emptyset$. Hence $D \setminus E$ is connected. □

3.6. REMOVABLE SINGULARITIES

A purely topological argument now yields the following corollary.

Corollary 3.6.4 *Every closed polar set E is totally disconnected.*

Proof. We need to show that if $w \in E$, then its component in E is just $\{w\}$. Without loss of generality, we may assume that $w = 0$. Let $\epsilon > 0$ and set

$$\Delta = \Delta(0,\epsilon), \quad \Delta^+ = \Delta \setminus [0,\epsilon), \quad \Delta^- = \Delta \setminus (-\epsilon, 0].$$

Choose $w_1, w_2 \in \Delta \setminus E$ with $\text{Im}(w_1) > 0$ and $\text{Im}(w_2) < 0$. By Theorem 3.6.3 both $\Delta^+ \setminus E$ and $\Delta^- \setminus E$ are connected, so we can join w_1 to w_2 by a path γ^+ in $\Delta^+ \setminus E$, and w_2 to w_1 by a path γ^- in $\Delta^- \setminus E$. Then $\gamma := \gamma^+ \cup \gamma^-$ is a closed path in $\Delta \setminus E$ which winds once around 0. It must therefore also wind once around every point in the same component of E as 0. Hence this component lies inside the disc $\Delta(0,\epsilon)$, and as ϵ is arbitrary, the component is just $\{0\}$. □

Here is a beautiful application of these ideas to complex analysis.

Theorem 3.6.5 (Radó–Stout Theorem) *Let D be a domain in $\mathbb{C}$, let E be a closed polar set, and let $f: D \to \mathbb{C}$ be a continuous function which is holomorphic on $D \setminus f^{-1}(E)$. Then f is holomorphic on the whole of D.*

Proof. If $f(D) \subset E$ then, as $f(D)$ is connected and E is totally disconnected, it follows that f is constant, in which case the result is obvious. From now on, we suppose that $f(D) \not\subset E$.

By Corollary 3.5.4 there exists a subharmonic function u on $\mathbb{C}$ such that $E = \{z : u(z) = -\infty\}$. Then $u \circ f$ is subharmonic on $D \setminus f^{-1}(E)$, and equals $-\infty$ on $f^{-1}(E)$, so it is subharmonic on the whole of D. Also $u \circ f \not\equiv -\infty$ on D, so by Theorem 3.5.1 $f^{-1}(E)$ is polar. We can now apply Corollary 3.6.2 to $\text{Re}\, f$ and $\text{Im}\, f$ to deduce that they are in fact harmonic on D, and hence that $f \in C^\infty(D)$. Since f satisfies the Cauchy–Riemann equations on $D \setminus E$, by continuity it must also do so on E, and hence it is holomorphic on D. □

Corollary 3.6.6 *Let D be a domain in $\mathbb{C}$, let f be a non-constant holomorphic function on D, and let E be a polar set. Then $f^{-1}(E)$ is also polar.*

Proof. If E is closed in $\mathbb{C}$, then this is a direct consequence of the previous proof. For the general case, it is sufficient to show that every compact subset of $f^{-1}(E)$ is polar, and this is easily deduced from the case already proved. □

In particular, it follows that the property of being a polar set is invariant under conformal mapping. Thus we can extend the notion of polarity to $\mathbb{C}_\infty$, by declaring a set E in $\mathbb{C}_\infty$ to be polar if $\phi(E)$ is polar for some conformal mapping ϕ of a neighbourhood of E into $\mathbb{C}$. It is easy to see that in fact E is polar in this sense if and only if $E \setminus \{\infty\}$ is polar in the standard sense.

Both the Liouville theorem (Corollary 2.3.4) and the maximum principle (Theorem 2.3.1) have extended versions, which will later prove very important.

Theorem 3.6.7 (Extended Liouville Theorem) *Let E be a closed polar subset of $\mathbb{C}$, and let u be a subharmonic function on $\mathbb{C} \setminus E$ which is bounded above. Then u is constant.*

Proof. By Theorem 3.6.1 u extends to be subharmonic on the whole of $\mathbb{C}$. Moreover, if $M = \sup_{\mathbb{C} \setminus E} u$, then $\max(u, M) = M$ on $\mathbb{C} \setminus E$, and hence everywhere on $\mathbb{C}$ by Theorem 2.7.5. Therefore u is bounded above on $\mathbb{C}$, and so applying Corollary 2.3.4 we deduce that u is constant. □

This theorem has a converse—see Exercise 2 below. It also has a version for holomorphic functions.

Corollary 3.6.8 *Let E be a closed polar subset of $\mathbb{C}$, and let f be a holomorphic function on $\mathbb{C} \setminus E$ such that $\mathbb{C} \setminus f(\mathbb{C} \setminus E)$ is non-polar. Then f is constant.*

Proof. Choose a compact non-polar set K such that $f(\mathbb{C} \setminus E) \subset \mathbb{C} \setminus K$, and let ν be an equilibrium measure for K. Then p_ν is harmonic and bounded below on $\mathbb{C} \setminus K$, so $-p_\nu \circ f$ is harmonic and bounded above on $\mathbb{C} \setminus E$. Hence by Theorem 3.6.7 $-p_\nu \circ f$ is constant. As $\lim_{z \to \infty} p_\nu(z) = \infty$, this implies that f is bounded on $\mathbb{C} \setminus E$. Applying Theorem 3.6.7 again, this time to $\operatorname{Re} f$ and $\operatorname{Im} f$, we deduce that f is constant. □

Theorem 3.6.9 (Extended Maximum Principle) *Let D be a domain in $\mathbb{C}$, and let u be a subharmonic function on D which is bounded above.*

(a) *If ∂D is polar, then u is constant.*

(b) *If ∂D is non-polar and $\limsup_{z \to \zeta} u(z) \leq 0$ for n.e. $\zeta \in \partial D$, then $u \leq 0$ on D.*

Proof. (a) Put $E = \partial D \setminus \{\infty\}$. Then E is a closed polar subset of $\mathbb{C}$, so by Theorem 3.6.3 $\mathbb{C} \setminus E$ is connected. Since D is a component of $\mathbb{C} \setminus E$, it follows that in fact $D = \mathbb{C} \setminus E$. The result is now an immediate consequence of Theorem 3.6.7.

(b) Given $\epsilon > 0$, define
$$E_\epsilon = \{\zeta \in \partial D \setminus \{\infty\} : \limsup_{z \to \zeta} u(z) \geq \epsilon\}.$$
Then E_ϵ is a closed polar subset of $\mathbb{C}$. Define v on $\mathbb{C} \setminus E_\epsilon$ by
$$v = \begin{cases} \max(u, \epsilon) & \text{on } D, \\ \epsilon & \text{on } \mathbb{C} \setminus (D \cup E_\epsilon). \end{cases}$$
By the gluing theorem (Theorem 2.4.5) v is subharmonic on $\mathbb{C} \setminus E_\epsilon$, and it is clearly bounded above there, so by Theorem 3.6.7 it is constant. Since $v = \epsilon$ on $\partial D \setminus (E_\epsilon \cup \{\infty\})$, which is non-empty, it follows that $v \equiv \epsilon$. Hence $u \leq \epsilon$ on D, and letting $\epsilon \to 0$ we deduce that $u \leq 0$ on D. $\square$

Exercises 3.6

1. Let f be holomorphic on a domain D, let ζ_0 be a non-isolated point of $\partial D \setminus \{\infty\}$, and suppose that $\lim_{z \to \zeta} f(z) = 0$ for all $\zeta \in \partial D$ sufficiently close to ζ_0. Use the Radó–Stout theorem to show that f extends holomorphically to a neighbourhood of ζ_0, and hence deduce that $f \equiv 0$ on D.

2. Let E be a closed subset of $\mathbb{C}$ with the property that every subharmonic function bounded above on $\mathbb{C} \setminus E$ is constant. Prove that E is polar. [Hint: If not, then E contains a non-polar compact set K. Let ν be an equilibrium measure for K, and consider $-p_\nu$.]

3.7 The Generalized Laplacian

By Theorem 2.4.4 a C^2 subharmonic function u satisfies $\Delta u \geq 0$. In this section we shall develop an appropriate generalization of this fact to arbitrary subharmonic functions. This turns out to be an important idea, with many applications.

Let D be a domain in $\mathbb{C}$. We denote by $C_c^\infty(D)$ the space of all C^∞ functions $\phi: D \to \mathbb{R}$ whose support $\operatorname{supp} \phi$ is a compact subset of D. If u is a C^2 subharmonic function on D, then, identifying Δu with the positive measure $\Delta u \, dA$, it follows from Green's theorem that

(3.2) $$\int_D \phi \Delta u = \int_D u \Delta \phi \, dA \quad (\phi \in C_c^\infty(D)).$$

Now if u is an arbitrary subharmonic function on D, with $u \not\equiv -\infty$, then by Theorem 2.5.1 u is locally integrable, and so the right-hand side of (3.2) still makes sense. We can therefore use it to *define* the left-hand side.

Definition 3.7.1 Let u be a subharmonic function on a domain D in $\mathbb{C}$, with $u \not\equiv -\infty$. The *generalized Laplacian* of u is the Radon measure Δu on D such that (3.2) holds.

(For details concerning Radon measures, in particular the Riesz representation theorem, see Section A.3 of the Appendix.)

To justify this definition, we need to prove the following theorem.

Theorem 3.7.2 *With the notation of Definition 3.7.1, Δu exists and is unique.*

The proof rests on a simple approximation lemma. We write $C_c(D)$ for the space of all continuous functions $\phi: D \to \mathbb{R}$ whose support $\operatorname{supp} \phi$ is a compact subset of D. Also, we define the sup-norm on $C_c(D)$ by

$$\|\phi\|_\infty := \sup_D |\phi| \quad (\phi \in C_c(D)).$$

Lemma 3.7.3 *Let $\phi \in C_c(D)$, and let U be a relatively compact open subset of D such that $\operatorname{supp} \phi \subset U$. Then there exist $(\phi_n)_{n \geq 1} \in C_c^\infty(D)$ such that $\operatorname{supp} \phi_n \subset U$ for all n and $\|\phi_n - \phi\|_\infty \to 0$. Moreover, if $\phi \geq 0$, then the (ϕ_n) can be chosen so that also $\phi_n \geq 0$ for each n.*

Proof. Extend ϕ to the whole $\mathbb{C}$ by defining $\phi \equiv 0$ on $\mathbb{C} \setminus D$. Then if $(\chi_r)_{r>0}$ are the functions used in Theorem 2.7.2, we have $\phi * \chi_r \in C^\infty(\mathbb{C})$ for all $r > 0$. Also

$$\operatorname{supp}(\phi * \chi_r) \subset \{z \in \mathbb{C} : \operatorname{dist}(z, \operatorname{supp} \phi) \leq r\},$$

so that $\operatorname{supp}(\phi * \chi_r) \subset U$ provided that r is sufficiently small. Finally,

$$\begin{aligned}\|\phi * \chi_r - \phi\|_\infty &= \sup_{z \in \mathbb{C}} \left| \int_{\Delta(0,r)} (\phi(z-w) - \phi(z)) \chi_r(w) \, dA(w) \right| \\ &\leq \sup_{\substack{z \in \mathbb{C} \\ |w| < r}} |\phi(z-w) - \phi(z)|,\end{aligned}$$

which tends to 0 as $r \to 0$ because ϕ is uniformly continuous on $\mathbb{C}$. Hence we may take $\phi_n = \phi * \chi_{\delta/n}$ ($n \geq 1$), where $\delta > 0$ is chosen sufficiently small. Also, with this definition, it is clear that if $\phi \geq 0$ then $\phi_n \geq 0$ for all n. □

Proof of Theorem 3.7.2. We begin with uniqueness. Suppose that μ_1 and μ_2 are Radon measures on D such that

$$\int_D \phi \, d\mu_1 = \int_D \phi \, d\mu_2 \quad (\phi \in C_c^\infty(D)).$$

3.7. THE GENERALIZED LAPLACIAN

Then using Lemma 3.7.3 it follows that this equation also holds for all $\phi \in C_c(D)$. By the uniqueness part of the Riesz representation theorem (Theorem A.3.2 in the Appendix), this implies that $\mu_1 = \mu_2$.

Now we turn to the question of existence. Define $\Lambda \colon C_c^\infty(D) \to \mathbb{R}$ by

$$\Lambda(\phi) = \int_D u \Delta\phi \, dA \quad (\phi \in C_c^\infty(D)).$$

Clearly Λ is a linear functional, and our first step is to show that it is positive, i.e. that

$$\phi \geq 0 \Rightarrow \Lambda(\phi) \geq 0.$$

Suppose then that $\phi \in C_c^\infty(D)$ with $\phi \geq 0$. Choose a relatively compact open subset U of D such that $\operatorname{supp}\phi \subset U$. By Corollary 2.7.3 there exist C^∞ subharmonic functions $(u_n)_{n \geq 1}$ on U such that $u_n \downarrow u$ there. By Theorem 2.4.4 $\Delta u_n \geq 0$ for each n, and so using Green's theorem it follows that

$$\int_D u_n \Delta\phi \, dA = \int_D \phi \Delta u_n \, dA \geq 0.$$

Letting $n \to \infty$ and applying the dominated convergence theorem, we deduce that

$$\int_D u \Delta\phi \, dA \geq 0,$$

or in other words $\Lambda(\phi) \geq 0$. Thus Λ is indeed positive.

Next we show that, given a relatively compact open subset V of D, there exists a constant C such that

(3.3) $\qquad |\Lambda(\phi)| \leq C\|\phi\|_\infty \quad (\phi \in C_c^\infty(D), \ \operatorname{supp}\phi \subset V).$

To do this, take $\psi \in C_c^\infty(D)$ such that $0 \leq \psi \leq 1$ and $\psi \equiv 1$ on V. Then, given $\phi \in C_c^\infty(D)$ with $\operatorname{supp}\phi \subset V$, we have

$$-\|\phi\|_\infty \psi \leq \phi \leq \|\phi\|_\infty \psi \quad \text{on } D,$$

so, since Λ is positive, it follows that

$$-\|\phi\|_\infty \Lambda(\psi) \leq \Lambda(\phi) \leq \|\phi\|_\infty \Lambda(\psi).$$

Thus (3.3) holds with $C = \Lambda(\psi)$.

Now combining (3.3) with Lemma 3.7.3, we deduce that Λ extends to a positive linear functional on the whole of $C_c(D)$. Therefore, by the existence part of the Riesz representation theorem, there is a Radon measure μ on D such that

$$\Lambda(\phi) = \int_D \phi \, d\mu \quad (\phi \in C_c(D)).$$

In particular,
$$\int_D u\Delta\phi\, dA = \int_D \phi\, d\mu \quad (\phi \in C_c^\infty(D)),$$
which completes the proof of existence. □

The reader familiar with distribution theory will recognize the generalized Laplacian as being just the Laplacian interpreted in the distributional sense. Although no previous knowledge of distribution theory is assumed in this book, it is helpful in understanding several of the results. For example, the potential p_μ can be regarded as the distributional convolution of the measure μ with the locally integrable function $\log|z|$, and the latter is just (a multiple of) the fundamental solution of the Laplacian. One might therefore expect Δp_μ to be the convolution of μ with a delta-function, i.e. a multiple of μ itself. That this is indeed the case is confirmed by the next theorem.

Theorem 3.7.4 *Let μ be a finite Borel measure on $\mathbb{C}$ with compact support. Then*
$$\Delta p_\mu = 2\pi\mu.$$

Proof. Given $\phi \in C_c^\infty(\mathbb{C})$, we have
$$\begin{aligned}
\int_{\mathbb{C}} p_\mu \Delta\phi\, dA &= \int_{\mathbb{C}} \left(\int_{\mathbb{C}} \log|z-w|\, d\mu(w) \right) \Delta\phi(z)\, dA(z) \\
&= \int_{\mathbb{C}} \left(\int_{\mathbb{C}} \log|z-w|\Delta\phi(z)\, dA(z) \right) d\mu(w).
\end{aligned}$$

(The use of Fubini's theorem is justified, because $\Delta\phi$ is bounded with compact support, and $\log|z|$ is locally integrable with respect to Lebesgue measure on $\mathbb{C}$.)

Now if $w \in \mathbb{C}$, then using Green's theorem we get
$$\begin{aligned}
&\int_{\mathbb{C}} \log|z-w|\Delta\phi(z)\, dA(z) \\
&= \lim_{\epsilon \to 0} \int_{|z-w|>\epsilon} \log|z-w|\Delta\phi(z)\, dA(z) \\
&= \lim_{\epsilon \to 0} \int_0^{2\pi} \left(\phi(w+re^{it}) - r\log r \frac{\partial\phi}{\partial r}(w+re^{it}) \right) \bigg|_{r=\epsilon} dt \\
&= 2\pi\phi(w).
\end{aligned}$$

Hence
$$\int_{\mathbb{C}} p_\mu \Delta\phi\, dA = \int_{\mathbb{C}} 2\pi\phi\, d\mu \quad (\phi \in C_c^\infty(\mathbb{C})),$$
which is what had to be proved. □

3.7. THE GENERALIZED LAPLACIAN

Corollary 3.7.5 *Let μ_1 and μ_2 be finite Borel measures on $\mathbb{C}$ with compact support. If $p_{\mu_1} = p_{\mu_2} + h$ on an open set U, where h is harmonic on U, then $\mu_1|_U = \mu_2|_U$.*

Proof. Since h is harmonic on U, we have

$$(\Delta p_{\mu_1})|_U = (\Delta p_{\mu_2})|_U.$$

The result therefore follows from Theorem 3.7.4. □

As an application of this result, we can justify the statement made in Section 3.3 concerning the uniqueness of equilibrium measures.

Theorem 3.7.6 *Let K be a compact non-polar subset of $\mathbb{C}$. Then its equilibrium measure ν is unique, and $\operatorname{supp} \nu \subset \partial_e K$ (the exterior boundary of K).*

Proof. We begin by observing that $\partial_e K$ is necessarily non-polar. For if it were polar, then by Theorem 3.6.3 $\mathbb{C} \setminus \partial_e K$ would be connected, and this would imply that $\partial_e K = K$, contrary to the assumption that K is non-polar.

Let ν and ν' be equilibrium measures for K and $\partial_e K$ respectively. It is sufficient to prove that $\nu = \nu'$. By Frostman's theorem (Theorem 3.3.4) $p_\nu \geq I(\nu)$ on $\mathbb{C}$ and $p_\nu = I(\nu)$ n.e. on K. Also p_ν is bounded above on each bounded component of $\mathbb{C} \setminus K$, so applying the extended maximum principle (Theorem 3.6.9) we deduce that $p_\nu \equiv I(\nu)$ there. Similarly, $p_{\nu'} \geq I(\nu')$ on $\mathbb{C}$ and $p_{\nu'} = I(\nu')$ n.e. on $\partial_e K$, and also $p_{\nu'} \equiv I(\nu')$ on each bounded component of $\mathbb{C} \setminus \partial_e K$. Finally, on the unbounded component of $\mathbb{C} \setminus K$, which is the same as the unbounded component of $\mathbb{C} \setminus \partial_e K$, the difference $(p_\nu - p_{\nu'})$ is harmonic and bounded, and so by the extended maximum principle again, $p_\nu - p_{\nu'} \equiv I(\nu) - I(\nu')$ there. Moreover, since

$$p_\nu(z) - p_{\nu'}(z) = (\log|z| + o(1)) - (\log|z| + o(1)) = o(1) \quad \text{as } z \to \infty,$$

it follows that $I(\nu) = I(\nu')$. Thus $p_\nu = p_{\nu'}$ n.e. on $\mathbb{C}$, hence a.e. on $\mathbb{C}$, and therefore everywhere on $\mathbb{C}$ by the weak identity principle (Theorem 2.7.5). Applying Corollary 3.7.5 we deduce that $\nu = \nu'$, as required. □

Corollary 3.7.7 *The equilibrium measure of a closed disc $\overline{\Delta}$ is normalized Lebesgue measure on $\partial \Delta$.*

Proof. By Theorem 3.7.6, the equilibrium measure is supported on $\partial \Delta$, and since it is unique it must be rotation-invariant. This implies that it is a multiple of Lebesgue measure on $\partial \Delta$. □

As a further application of Theorem 3.7.4 we can compute $\Delta(\log|f|)$ when f is holomorphic.

Theorem 3.7.8 *Let f be holomorphic on a domain D, with $f \not\equiv 0$. Then $\Delta(\log|f|)$ consists of (2π)-masses at the zeros of f, counted according to multiplicity.*

Proof. Given a relatively compact open subset U of D, we can write
$$f(z) = (z - w_1) \cdots (z - w_n)g(z) \quad (z \in U),$$
where $w_1, \ldots, w_n$ are the zeros of f in U, and g is holomorphic and non-zero on U. Then
$$\log|f(z)| = \sum_1^n \log|z - w_j| + \log|g(z)| = p_\mu(z) + h(z) \quad (z \in U),$$
where μ consists of unit masses at $w_1, \ldots, w_n$ and h is harmonic on U. By Theorem 3.7.4 $\Delta(\log|f|) = 2\pi\mu$ on U. As this holds for each such U, the result follows. □

The proof above shows that $\log|f|$ can be expressed locally as the sum of a potential and a harmonic function. This is actually a special case of a quite general result.

Theorem 3.7.9 (Riesz Decomposition Theorem) *Let u be a subharmonic function on a domain D in $\mathbb{C}$, with $u \not\equiv -\infty$. Then, given a relatively compact open subset U of D, we can decompose u as*
$$u = p_\mu + h \quad on\ U,$$
where $\mu = (2\pi)^{-1}\Delta u|_U$ and h is harmonic on U.

This is a very powerful theorem. It means that many problems about general subharmonic functions can be reduced to questions about potentials. For example, Exercise 3.5.1, which appeared to be a special case of Theorem 3.5.1, turns out not to be so special after all. Another application will appear in the next section.

Most of the work for Theorem 3.7.9 has already been done. What remains to be proved is the following lemma, which is a converse to Corollary 3.7.5.

Lemma 3.7.10 (Weyl's Lemma) *Let u and v be subharmonic functions on a domain D in $\mathbb{C}$, with $u, v \not\equiv -\infty$. If $\Delta u = \Delta v$, then $u = v + h$ where h is harmonic on D.*

3.7. THE GENERALIZED LAPLACIAN

Proof. Let $(\chi_r)_{r>0}$ be the functions used in the smoothing theorem, Theorem 2.7.2, and for $r > 0$ write

$$D_r := \{z \in D : \operatorname{dist}(z, \partial D) > r\}.$$

Then $u * \chi_r \in C^\infty(D_r)$, and for $z \in D_r$ we have

$$\begin{aligned}\Delta(u * \chi_r)(z) &= \int u(w) \Delta_z \chi_r(z - w) \, dA(w) \\ &= \int u(w) \Delta_w \chi_r(z - w) \, dA(w) \\ &= \int \phi \Delta u,\end{aligned}$$

where $\phi(w) = \chi_r(z - w) \in C_c^\infty(D)$. The same calculation works with u replaced by v. Since $\Delta u = \Delta v$, it follows that $\Delta(u * \chi_r) = \Delta(v * \chi_r)$ on D_r. Therefore there exists a harmonic function h_r on D_r such that

$$u * \chi_r = v * \chi_r + h_r \quad \text{on } D_r.$$

Now by Theorem 2.7.2 applied to $\pm h_r$, we have $h_r * \chi_s = h_r$ on D_{r+s} for each $s > 0$, and hence

$$h_r = h_r * \chi_s = (u - v) * \chi_r * \chi_s = h_s * \chi_r = h_s \quad \text{on } D_{r+s}.$$

Therefore there is a single harmonic function h on D such that, for each $r > 0$,

$$u * \chi_r = v * \chi_r + h \quad \text{on } D_r.$$

Letting $r \downarrow 0$ and using Theorem 2.7.2, we deduce that $u = v + h$ on D. □

Proof of Theorem 3.7.9. Put $\mu = (2\pi)^{-1} \Delta u|_U$. Then by Theorem 3.7.4

$$\Delta p_\mu = 2\pi \mu = \Delta u \quad \text{on } U.$$

Applying Lemma 3.7.10 on each component of U, it follows that $u = p_\mu + h$ on U, where h is harmonic on U. □

Exercises 3.7

1. Show that $\Delta(\log^+ |z|)$ is Lebesgue measure on the unit circle.

2. Let D be a domain in $\mathbb{C}$, let $w \in D$, and let h be a positive harmonic function on $D \setminus \{w\}$. Show that $-h$ extends to be subharmonic on D, and deduce that there exist a harmonic function k on D and a constant $C \geq 0$ such that

$$h(z) = k(z) - C \log |z - w| \quad (z \in D \setminus \{w\}).$$

3. (i) Let μ be a finite Borel measure on $\Delta(0,1)$. Show that if $0 < r < 1$ and $\epsilon > 0$, then

$$\left| \frac{1}{2\pi} \int_0^{2\pi} p_\mu(re^{it}) \, dt \right| = \left| \int_{\Delta(0,1)} \max(\log r, \log|w|) \, d\mu(w) \right|$$
$$\leq |\log r| \Big(\mu(\Delta(0, r^\epsilon)) + \epsilon\mu(\Delta(0,1)) \Big).$$

Deduce that if $\mu(\{0\}) = 0$ then

$$\frac{1}{\log r} \left(\frac{1}{2\pi} \int_0^{2\pi} p_\mu(re^{it}) \, dt \right) \to 0 \quad \text{as } r \to 0.$$

(ii) Let u be a subharmonic function on $\Delta(0, \rho)$ such that $u(0) = -\infty$ and $u \not\equiv -\infty$. Prove that, in the notation of Definition 2.6.7,

$$\lim_{r \to 0} \frac{M_u(r)}{\log r} = \lim_{r \to 0} \frac{C_u(r)}{\log r} = \lim_{r \to 0} \frac{B_u(r)}{\log r} = \Delta u(\{0\}).$$

[Hint: First reduce to the case $\Delta u(\{0\}) = 0$. The result for C_u then follows from (i). For M_u and B_u use Theorem 2.6.8 (b) and Exercise 2.6.6 respectively.]

4. Let $(u_n)_{n \geq 1}$ be subharmonic functions on a domain D in $\mathbb{C}$, and suppose that $u_n \downarrow u$ on D, where $u \not\equiv -\infty$. Show that as $n \to \infty$

$$\int_D \phi \Delta u_n \to \int_D \phi \Delta u \quad (\phi \in C_c(D)).$$

[Hint: First prove the result for $\phi \in C_c^\infty(D)$. Use this to deduce that the measures $(\Delta u_n)_{n \geq 1}$ are uniformly bounded on each compact subset of D.]

3.8 Thinness

Let u be a function subharmonic on a neighbourhood of $\zeta \in \mathbb{C}$. Even though u may be discontinuous at ζ, it is still always true that

(3.4) $$\limsup_{\substack{z \to \zeta \\ z \neq \zeta}} u(z) = u(\zeta).$$

For by upper semicontinuity, we certainly have $\limsup_{z \to \zeta} u(z) \leq u(\zeta)$, and if the inequality were strict, then u would violate the submean inequality on small circles round ζ. Thus the value of u at ζ is completely determined by its values on a punctured disc around ζ. It turns out to be useful to know to what extent the punctured disc may be replaced by a smaller set S.

3.8. THINNESS

Definition 3.8.1 Let S be a subset of $\mathbb{C}$ and let $\zeta \in \mathbb{C}$. Then S is *non-thin* at ζ if $\zeta \in \overline{S \setminus \{\zeta\}}$ and if, for every subharmonic function u defined on a neighbourhood of ζ,
$$\limsup_{\substack{z \to \zeta \\ z \in S \setminus \{\zeta\}}} u(z) = u(\zeta).$$
Otherwise we say that S is *thin* at ζ.

A complete characterization of thinness is quite complicated, and must await developments in Chapter 5. However, for many purposes it is enough to be able to handle a few important special cases, which we shall study in this section.

We begin with some elementary remarks. Thinness is obviously a local property, i.e. S is non-thin at ζ if and only if $U \cap S$ is non-thin at ζ for each open neighbourhood U of ζ. It is also invariant under conformal mapping, so that although we have defined thinness in the plane, we could equally well study it on the sphere. Lastly, if two sets are both thin at a particular point, then so is their union—the proof is left as an easy exercise.

From (3.4) it follows that a set S is non-thin at each point of its interior. In particular, an open set is non-thin at every point of itself. On the other hand, it be can thin at some of its boundary points. For example, let u be a subharmonic function which is discontinuous at ζ, and choose α so that $\liminf_{z \to \zeta} u(z) < \alpha < u(\zeta)$. Then $S := \{z : u(z) < \alpha\}$ is an open set with $\zeta \in \partial S$, and clearly S is thin at ζ.

We now look at some special types of set S, beginning with small ones.

Theorem 3.8.2 *An F_σ polar set is thin at every point of $\mathbb{C}$.*

Proof. Let S be an F_σ polar set, and let $\zeta \in \mathbb{C}$. Then $S \setminus \{\zeta\}$ is also an F_σ polar set, and is obviously disjoint from $\{\zeta\}$, so by Theorem 3.5.2 there exists a subharmonic function u on $\mathbb{C}$ such that $u = -\infty$ on $S \setminus \{\zeta\}$ and $u(\zeta) > -\infty$. Therefore S is thin at ζ. □

At the other extreme we have the following theorem.

Theorem 3.8.3 *A connected set containing more than one point is non-thin at every point of its closure.*

The proof is based on a lemma which is actually a special case of the main result.

Lemma 3.8.4 *Let u be a subharmonic function on $\Delta(0,1)$. If $u \leq 0$ on the segment $(0,1)$, then $u(0) \leq 0$.*

Proof. Replacing u by $\max(u,0)$, we can suppose that $u \geq 0$ on $\Delta(0,1)$ and $u = 0$ on $(0,1)$. We then need to show that $u(0) = 0$.

Define v on $\Delta(0,1) \setminus \{0\}$ by
$$v(z) = \begin{cases} u(z^2), & \operatorname{Im} z > 0, \\ 0, & \operatorname{Im} z \leq 0. \end{cases}$$

Then v is subharmonic on $\Delta(0,1) \setminus \{0\}$ by the gluing theorem (Theorem 2.4.5). Also v is bounded above near 0, so by the removable singularity theorem (Theorem 3.6.1) it extends to a function subharmonic on the whole of $\Delta(0,1)$. Then by Theorem 2.6.8 (c),
$$v(0) = \lim_{r \to 0} M_v(r) = \lim_{r \to 0} M_u(r^2) = u(0),$$
and also
$$v(0) = \lim_{r \to 0} C_v(r) = \lim_{r \to 0} \frac{1}{2} C_u(r^2) = \frac{1}{2} u(0).$$
Combining the two yields $u(0) = 0$, as required. $\square$

Proof of Theorem 3.8.3. We argue by contradiction. Let S be a connected set with at least two points, and suppose, if possible, that S is thin at some point ζ of its closure. Applying a conformal mapping, we may assume that $\zeta = 0$. Then there exists a subharmonic function u, defined on a neighbourhood of 0, such that
$$\limsup_{\substack{z \to 0 \\ z \in S \setminus \{0\}}} u(z) < u(0).$$

By the Riesz decomposition theorem (Theorem 3.7.9) we can decompose u on a neighbourhood of 0 as $u = p_\mu + h$, where p_μ is the potential of a finite Borel measure μ of compact support, and h is harmonic. Since h is continuous, it follows that
$$\limsup_{\substack{z \to 0 \\ z \in S \setminus \{0\}}} p_\mu(z) < p_\mu(0).$$

Now define $T \colon \mathbb{C} \to \mathbb{R}$ by $T(z) = |z|$, and set
$$\mu_1(B) = \mu(T^{-1}(B)) \quad (\text{Borel } B \subset \mathbb{C}),$$
so that μ_1 is also a finite Borel measure with compact support. Then,
$$p_{\mu_1}(|z|) = \int \log\big||z| - |w|\big| \, d\mu(w) \leq p_\mu(z) \quad (z \in \mathbb{C}),$$
with equality if $z = 0$. Therefore
$$\limsup_{\substack{z \to 0 \\ z \in S \setminus \{0\}}} p_{\mu_1}(|z|) < p_{\mu_1}(0).$$

3.8. THINNESS

Since S is connected and contains a point other than 0, it follows that $\{|z| : z \in S\}$ includes an interval $(0, \alpha)$ for some $\alpha > 0$. Hence

$$\limsup_{\substack{z \to 0 \\ z \in (0,\alpha)}} p_{\mu_1}(z) < p_{\mu_1}(0).$$

It is therefore possible to choose constants r, s so that $u_1(z) := p_{\mu_1}(rz) + s$ is subharmonic on $\Delta(0,1)$, and satisfies $u_1 \leq 0$ on $(0,1)$ and $u_1(0) > 0$. This violates Lemma 3.8.4, and gives the desired contradiction. □

Combining the last two theorems leads immediately to a generalization of Corollary 3.6.4.

Corollary 3.8.5 *Every F_σ polar set is totally disconnected.* □

A set may be thin at 'many' points—as an extreme example, a countable dense subset of $\mathbf{C}$ is thin everywhere. However, as our final theorem of this section shows, a set cannot be thin at too many points of itself.

Theorem 3.8.6 *A subset S of $\mathbf{C}$ is non-thin at nearly every point of itself.*

In other words, there is a Borel polar set E such that S is non-thin at every point of $S \setminus E$.

Proof. Let (U_j) be a countable base of open sets for $\mathbf{C}$ with $\operatorname{diam}(U_j) < 1$. For each j, let $\mathcal{V}_j$ be the collection of all subharmonic functions v on U_j such that
$$\begin{aligned} v &\leq 0 && \text{on } U_j, \\ v &\leq -1 && \text{on } U_j \cap S. \end{aligned}$$

Set $u_j = \sup_{v \in \mathcal{V}_j} v$, and let u_j^* be its upper semicontinuous regularization. Then by the Brelot–Cartan theorem (Theorem 3.4.2) there is a Borel polar set E_j such that $u_j^* = u_j$ on $U_j \setminus E_j$. Set $E = \cup_j E_j$. Then E is a Borel polar set, and we shall show that S is non-thin at each point of $S \setminus E$.

Suppose that $\zeta \in S$ and that S is thin at ζ. If ζ is a non-isolated point of S, then there exists a subharmonic function u on a neighbourhood of ζ such that
$$\limsup_{\substack{z \to \zeta \\ z \in S \setminus \{\zeta\}}} u(z) < -1 < u(\zeta) < 0.$$

Therefore there is a neighbourhood of ζ, which we may take to be member U_j of the countable base, such that
$$\begin{aligned} u &\leq 0 && \text{on } U_j, \\ u &\leq -1 && \text{on } U_j \cap S \setminus \{\zeta\}. \end{aligned}$$

If ζ is an isolated point of S, then we reach the same conclusion by choosing U_j so that $U_j \cap S = \{\zeta\}$, and setting $u \equiv 0$. Then, for each $\epsilon > 0$, the function $v_\epsilon(z) := u(z) + \epsilon \log|z-\zeta|$ belongs to the class $\mathcal{V}_j$, and so $u_j \geq v_\epsilon$. Letting $\epsilon \to 0$, we deduce that $u_j \geq u$ on $U_j \setminus \{\zeta\}$, and hence that $u_j^* \geq u$ on U_j. In particular, $u_j^*(\zeta) \geq u(\zeta) > -1$. On the other hand it is clear that $u_j(\zeta) \leq -1$ since $\zeta \in S$. Hence $\zeta \in E_j$. We have therefore shown that the only points of S at which S can be thin are those that lie in E. This completes the proof. □

As a special case, we obtain a converse to Theorem 3.8.2.

Corollary 3.8.7 *A set S which is thin at every point of itself must be polar.* □

Exercises 3.8

1. Show that if two sets S_1 and S_2 are both thin at ζ, then so is $S_1 \cup S_2$.

2. Let U be an open subset of $\mathbb{C}$, and let S be a subset of U such that $U \setminus S$ has Lebesgue measure zero. Show that S is non-thin at every point of U.

3. By examining the proof of Theorem 3.8.3, show that if S is thin at ζ then there exist positive numbers $r_n \to 0$ such that $S \cap \partial \Delta(\zeta, r_n) = \emptyset$. Deduce that if u is subharmonic on a neighbourhood of ζ, then

$$\limsup_{r \to 0} \left(\inf_{|z-\zeta|=r} u(z) \right) = u(\zeta).$$

4. Use Theorem 3.8.6 to show that $\mathbb{R}$ is non-thin at each point of itself, and hence give another proof of Lemma 3.8.4.

Notes on Chapter 3

§3.2

Hausdorff measures can be used to provide a fairly accurate appraisal of whether or not a given compact set K is polar. In fact if $m_h(K) < \infty$, where

$$h(t) = \left(\log \frac{1}{t} \right)^{-1},$$

then K is polar, while if $m_h(K) > 0$, where

$$h(t) = \left(\log\frac{1}{t}\right)^{-1}\left(\log\log\frac{1}{t}\right)^{-1-\delta} \quad (\delta > 0),$$

then K is non-polar. However, no complete description of polar sets in terms of Hausdorff measures is possible. For more details, see [34, §5.4] and [22, Chapter IV].

Exercise 2, which is due to Bernstein, indicates that non-Borel polar sets are unlikely to be of much interest. A fuller discussion of this follows in the Notes on Section 5.1.

§3.4

Part (c) of Theorem 3.4.3 is often singled out under the name of Hartogs' lemma. Its multi-variable analogue is a key step in the proof of Hartogs' separate analyticity theorem (see e.g. [39, §2.4]).

§3.5

Deny's sharp converse to Theorem 3.5.1 appears in [24]. The proof given here of the rather simpler Lemma 3.5.3 is based on [66, Proposition 18.4].

§3.7

Exercise 2 is Bôcher's theorem. There are also more elementary proofs, not depending on the Riesz decomposition; see for instance [7, §3.9].

§3.8

Thinness can be characterized in terms of the so-called *fine topology*, namely the weakest topology on $\mathbf{C}$ with respect to which all subharmonic functions are continuous. In fact S is non-thin at ζ precisely when ζ is a fine limit point of S. Several aspects of potential theory are most naturally treated in the context of the fine topology, and a great deal is known about this topology. For more information we refer to Brelot's book [18].

Chapter 4

The Dirichlet Problem

4.1 Solution of the Dirichlet Problem

We recall from Definition 1.2.1 that, given a domain D and a continuous function $\phi: \partial D \to \mathbb{R}$, the *Dirichlet problem* is to find a harmonic function h on D such that $\lim_{z \to \zeta} h(z) = \phi(\zeta)$ for all $\zeta \in \partial D$. By Theorem 1.2.2, if such a solution h exists, then it is unique. Also, if D is a disc, then a solution always does exist, and Theorem 1.2.4 even gives a formula for it.

For a general domain D, the situation is more complicated. In this case, the Dirichlet problem, at least in the form stated above, may well have no solution. For example, take $D = \{z : 0 < |z| < 1\}$, and let $\phi: \partial D \to \mathbb{R}$ be given by

$$\phi(\zeta) = \begin{cases} 0, & |\zeta| = 1, \\ 1, & |\zeta| = 0. \end{cases}$$

Then by Corollary 3.6.2, any solution h would have a removable singularity at 0, and the maximum principle would then imply that $h(0) \leq 0$, violating the condition that $\lim_{z \to 0} h(z) = \phi(0) = 1$.

In this section and the next, we shall consider conditions under which a solution does exist, and also, even more importantly, derive a natural reformulation of the Dirichlet problem which *always* has a solution. To this end, it is convenient to extend the set-up described above in two ways.

Firstly, we shall allow D to be any proper subdomain of $\mathbb{C}_\infty$. Of course, since the Dirichlet problem is invariant under conformal mapping of the sphere, this is really no more general than working on a subdomain of $\mathbb{C}$. However, the gain in flexibility does turn out to be useful. We shall exploit without further comment the fact that harmonicity, subharmonicity and polarity all extend in a natural way to $\mathbb{C}_\infty$.

The other generalization will be to consider arbitrary bounded functions $\phi: \partial D \to \mathbb{R}$, rather than just continuous ones. Although certainly no solution to the Dirichlet problem is possible if ϕ is discontinuous, it is nevertheless useful to allow this extra freedom, as will become clear later.

The key idea, sometimes called the Perron method, is enshrined in the following definition.

Definition 4.1.1 Let D be a proper subdomain of $\mathbb{C}_\infty$, and let $\phi: \partial D \to \mathbb{R}$ be a bounded function. The associated *Perron function* $H_D \phi: D \to \mathbb{R}$ is defined by
$$H_D \phi = \sup_{u \in \mathcal{U}} u,$$
where $\mathcal{U}$ denotes the family of all subharmonic functions u on D such that $\limsup_{z \to \zeta} u(z) \leq \phi(\zeta)$ for each $\zeta \in \partial D$.

The motivation for this definition is that, if the Dirichlet problem has a solution at all, then $H_D \phi$ is it! Indeed, if h is such a solution, then certainly $h \in \mathcal{U}$, and so $h \leq H_D \phi$. On the other hand, by the maximum principle, if $u \in \mathcal{U}$ then $u \leq h$ on D, and so $H_D \phi \leq h$. Therefore $H_D \phi = h$.

Our first result is that, regardless of whether the Dirichlet problem has a solution, $H_D \phi$ is always a bounded harmonic function.

Theorem 4.1.2 *Let D be a proper subdomain of $\mathbb{C}_\infty$, and let $\phi: \partial D \to \mathbb{R}$ be a bounded function. Then $H_D \phi$ is harmonic on D, and*

(4.1) $$\sup_D |H_D \phi| \leq \sup_{\partial D} |\phi|.$$

The proof hinges on the following lemma.

Lemma 4.1.3 (Poisson Modification) *Let D be a domain in $\mathbb{C}$, let Δ be an open disc with $\overline{\Delta} \subset D$, and let u be a subharmonic function on D with $u \not\equiv -\infty$. If we define $\tilde{u}$ on D by*
$$\tilde{u} = \begin{cases} P_\Delta u & \text{on } \Delta, \\ u & \text{on } D \setminus \Delta, \end{cases}$$
then $\tilde{u}$ is subharmonic on D, harmonic on Δ, and $\tilde{u} \geq u$ on D.

Proof. First note that Corollary 2.5.2 guarantees that u is Lebesgue integrable on $\partial \Delta$, so $P_\Delta u$ makes sense. Theorem 1.2.4 (a) then tells us that $P_\Delta u$ is harmonic on Δ, and by Theorem 2.4.1 (b) $P_\Delta u \geq u$ there. It thus remains to show that $\tilde{u}$ is subharmonic on D, and by the gluing theorem (Theorem 2.4.5) this will follow provided that
$$\limsup_{z \to \zeta} P_\Delta u(z) \leq u(\zeta) \quad (\zeta \in \partial \Delta).$$

4.1. SOLUTION OF THE DIRICHLET PROBLEM

To prove this inequality, choose continuous functions ψ_n on $\partial\Delta$ such that $\psi_n \downarrow u$ there. Then, using Theorem 1.2.4 (b), we have

$$\limsup_{z \to \zeta} P_\Delta u(z) \leq \lim_{z \to \zeta} P_\Delta \psi_n(z) = \psi_n(\zeta) \quad (\zeta \in \partial\Delta),$$

and the desired conclusion follows by letting $n \to \infty$. □

Proof of Theorem 4.1.2. By applying a conformal mapping of the sphere, we can suppose that D is a subdomain of $\mathbb{C}$.

Let $\mathcal{U}$ be the family defined in Definition 4.1.1. If we set $M = \sup_{\partial D}|\phi|$, then certainly $-M \in \mathcal{U}$, so $H_D\phi \geq -M$. Also, given $u \in \mathcal{U}$, it follows from the maximum principle that $u \leq M$ on D, and therefore $H_D\phi \leq M$. This proves (4.1).

To show that $H_D\phi$ is harmonic on D, it suffices to prove harmonicity on each open disc Δ with $\overline{\Delta} \subset D$. Fix such a Δ, and also a point $w_0 \in \Delta$. By definition of $H_D\phi$, we can find $(u_n)_{n\geq 1} \in \mathcal{U}$ such that $u_n(w_0) \to H_D\phi(w_0)$. Replacing u_n by $\max(u_1, \ldots, u_n)$, we can further suppose that $u_1 \leq u_2 \leq u_3 \leq \cdots$ on D. Now for each n, let $\widetilde{u}_n$ denote the Poisson modification of u_n, as defined in Lemma 4.1.3. Then we also have $\widetilde{u}_1 \leq \widetilde{u}_2 \leq \widetilde{u}_3 \leq \cdots$ on D, and we claim that $\widetilde{u} := \lim_{n\to\infty} \widetilde{u}_n$ satisfies:

(i) $\widetilde{u} \leq H_D\phi$ on D;

(ii) $\widetilde{u}(w_0) = H_D\phi(w_0)$;

(iii) $\widetilde{u}$ is harmonic on Δ.

Indeed, by Lemma 4.1.3 each $\widetilde{u}_n$ is subharmonic on D, and evidently

$$\limsup_{z \to \zeta} \widetilde{u}_n(z) = \limsup_{z \to \zeta} u_n(z) \leq \phi(\zeta) \quad (\zeta \in \partial D),$$

so that $\widetilde{u}_n \in \mathcal{U}$. Hence $\widetilde{u}_n \leq H_D\phi$ for all n, and therefore $\widetilde{u} \leq H_D\phi$, which gives (i). Lemma 4.1.3 also tells us that $\widetilde{u}_n \geq u_n$, so

$$\widetilde{u}(w_0) = \lim_{n\to\infty} \widetilde{u}_n(w_0) \geq \lim_{n\to\infty} u_n(w_0) = H_D\phi(w_0).$$

This proves (ii), since the reverse inequality follows from (i). Finally, each $\widetilde{u}_n$ is harmonic on Δ, so by Harnack's theorem (Theorem 1.3.9) the same is true of $\widetilde{u}$, which gives (iii). The theorem would therefore be proved if we could show that $\widetilde{u} = H_D\phi$ on Δ.

To this end, take an arbitrary point $w \in \Delta$, and choose $(v_n)_{n\geq 1} \in \mathcal{U}$ such that $v_n(w) \to H_D\phi(w)$. Replacing v_n by $\max(u_1, \ldots, u_n, v_1, \ldots, v_n)$, we can suppose that $v_1 \leq v_2 \leq v_3 \leq \cdots$ and $v_n \geq u_n$ on D. Let $\widetilde{v}_n$ denote the Poisson modification of v_n. Then as before $\widetilde{v}_n \uparrow \widetilde{v}$, where:

(i) $\tilde{v} \leq H_D\phi$ on D;

(ii) $\tilde{v}(w) = H_D\phi(w)$;

(iii) $\tilde{v}$ is harmonic on Δ.

In particular, (i) implies that
$$\tilde{v}(w_0) \leq H_D\phi(w_0) = \tilde{u}(w_0).$$
On the other hand, $\tilde{v}_n \geq \tilde{u}_n$ for each n, so $\tilde{v} \geq \tilde{u}$. Thus the function $\tilde{u} - \tilde{v}$, which is harmonic on Δ, attains a maximum value of 0 at w_0. By the maximum principle, this implies that $\tilde{u} - \tilde{v} \equiv 0$ on Δ. In particular, it follows that
$$\tilde{u}(w) = \tilde{v}(w) = H_D\phi(w).$$
Since w is arbitrary, we have shown that $\tilde{u} = H_D\phi$ on Δ, as required. $\square$

From the definition of $H_D\phi$, one might expect that $\lim_{z\to\zeta} H_D\phi(z) = \phi(\zeta)$ at each $\zeta \in \partial D$. But if $D = \{z : 0 < |z| < 1\}$ then this cannot be true, because, as we have seen, the Dirichlet problem may have no solution. It is instructive to see exactly what goes wrong.

First let
$$\phi(\zeta) = \begin{cases} 0, & |\zeta| = 1, \\ 1, & |\zeta| = 0. \end{cases}$$
If $u \in \mathcal{U}$, then by the extended maximum principle (Theorem 3.6.9) $u \leq 0$ on D, and so $H_D\phi \leq 0$. Since $0 \in \mathcal{U}$, in fact $H_D\phi \equiv 0$ on D.

Now let
$$\phi(\zeta) = \begin{cases} 0, & |\zeta| = 1, \\ -1, & |\zeta| = 0. \end{cases}$$
The same argument as before (even using the ordinary maximum principle) shows that $H_D\phi \leq 0$. This time $0 \notin \mathcal{U}$. However, it is true that $\epsilon \log|z| \in \mathcal{U}$ for each $\epsilon > 0$, and so once again $H_D\phi \equiv 0$ on D.

In both cases, the isolated boundary point 0 lacked sufficient 'influence' on the subharmonic functions in $\mathcal{U}$, and the result was that $H_D\phi$ had the wrong boundary limit there. To overcome this problem, we introduce the notion of a barrier.

Definition 4.1.4 Let D be a proper subdomain of $\mathbb{C}_\infty$, and let $\zeta_0 \in \partial D$. A *barrier* at ζ_0 is a subharmonic function b defined on $D \cap N$, where N is an open neighbourhood of ζ_0, satisfying
$$b < 0 \text{ on } D \cap N \quad \text{and} \quad \lim_{z\to\zeta_0} b(z) = 0.$$

A boundary point at which a barrier exists is called *regular*, otherwise it is *irregular*. If every $\zeta \in \partial D$ is regular, then D is called a *regular domain*.

4.1. SOLUTION OF THE DIRICHLET PROBLEM

Theorem 4.1.5 *Let D be a proper subdomain of $\mathbb{C}_\infty$, and let ζ_0 be a regular boundary point of D. If $\phi: \partial D \to \mathbb{R}$ is a bounded function which is continuous at ζ_0, then*

$$\lim_{z \to \zeta_0} H_D\phi(z) = \phi(\zeta_0).$$

This time we need two lemmas. The first is a simple consequence of Definition 4.1.1.

Lemma 4.1.6 *If D is a proper subdomain of $\mathbb{C}_\infty$ and $\phi: \partial D \to \mathbb{R}$ is a bounded function, then*

$$H_D\phi \leq -H_D(-\phi) \quad \text{on } D.$$

Proof. Let $\mathcal{U}$ be the family of subharmonic functions prescribed in Definition 4.1.1, and let $\mathcal{V}$ be the corresponding family for $-\phi$. Then, given $u \in \mathcal{U}$ and $v \in \mathcal{V}$, their sum is subharmonic on D and satisfies

$$\limsup_{z \to \zeta}(u+v)(z) \leq \phi(\zeta) - \phi(\zeta) = 0 \quad (\zeta \in \partial D).$$

Hence by the maximum principle $u + v \leq 0$ on D. Taking suprema over all such u and v, we get $H_D\phi + H_D(-\phi) \leq 0$ on D, which gives the result. □

The second lemma enables us to 'globalize' barriers.

Lemma 4.1.7 (Bouligand's Lemma) *Let ζ_0 be a regular boundary point of a domain D, and let N_0 be an open neighbourhood of ζ_0. Then, given $\epsilon > 0$, there exists a subharmonic function b_ϵ on D such that*

$$b_\epsilon < 0 \text{ on } D, \quad b_\epsilon \leq -1 \text{ on } D \setminus N_0, \quad \text{and} \quad \liminf_{z \to \zeta_0} b_\epsilon(z) \geq -\epsilon.$$

Proof. We may suppose that $\zeta_0 \neq \infty$ (otherwise apply a conformal mapping). Since ζ_0 is regular, there exists a neighbourhood N of ζ_0 and a barrier b on $D \cap N$ as in Definition 4.1.4.

Let $\Delta = \Delta(\zeta_0, \rho)$, where ρ is chosen small enough that $\overline{\Delta} \subset N \cap N_0$. Then the normalized Lebesgue measure on $\partial \Delta$ is a regular measure (see Theorem A.2.2 in the Appendix), so we can find a compact set $K \subset D \cap \partial \Delta$ such that $L := (D \cap \partial \Delta) \setminus K$ has measure $< \epsilon$. Since L is open in $\partial \Delta$, it follows from Theorem 1.2.4 (b) that

$$\lim_{\substack{z \to \eta \\ z \in D}} P_\Delta 1_L(z) = 1 \quad (\eta \in L).$$

Now put $m = -\sup_K b$, so that $m > 0$. Then for $\eta \in D \cap \partial\Delta$,

$$\limsup_{\substack{z \to \eta \\ z \in D \cap \Delta}} \left(\frac{b(z)}{m} - P_\Delta 1_L(z) \right) \le \left\{ \begin{array}{cc} b(\eta)/m - 0, & \text{if } \eta \in K \\ 0 - 1, & \text{if } \eta \in L \end{array} \right\} \le -1.$$

Hence if we define b_ϵ on D by

$$b_\epsilon = \left\{ \begin{array}{ll} \max(-1, (b/m - P_\Delta 1_L)) & \text{on } D \cap \Delta, \\ -1 & \text{on } D \setminus \Delta, \end{array} \right.$$

then by the gluing theorem (Theorem 2.4.5) b_ϵ is subharmonic on D. Clearly

$$b_\epsilon < 0 \text{ on } D \quad \text{and} \quad b_\epsilon \le -1 \text{ on } D \setminus N_0.$$

Finally, we have

$$\liminf_{z \to \zeta_0} b_\epsilon(z) \ge \lim_{z \to \zeta_0} \left(\frac{b(z)}{m} - P_\Delta 1_L(z) \right) = 0 - P_\Delta 1_L(\zeta_0) > -\epsilon,$$

the last inequality arising from the fact that, as ζ_0 is the centre of Δ, the value of $P_\Delta 1_L(\zeta_0)$ is exactly the normalized Lebesgue measure of L. □

Proof of Theorem 4.1.5. Let $\epsilon > 0$. Since ϕ is continuous at ζ_0, there exists an open neighbourhood N_0 of ζ_0 such that

$$\zeta \in \partial D \cap \overline{N}_0 \Rightarrow |\phi(\zeta) - \phi(\zeta_0)| < \epsilon.$$

Construct b_ϵ as in Lemma 4.1.7, and set

$$u = \phi(\zeta_0) - \epsilon + (M + \phi(\zeta_0)) b_\epsilon,$$

where $M = \sup_{\partial D} |\phi|$. Then u is subharmonic on D, and if $\zeta \in \partial D$ then

$$\limsup_{z \to \zeta} u(z) \le \left\{ \begin{array}{ll} \phi(\zeta_0) - \epsilon + 0, & \text{if } \zeta \in \partial D \cap \overline{N}_0 \\ \phi(\zeta_0) - \epsilon - (M + \phi(\zeta_0)), & \text{if } \zeta \in \partial D \setminus \overline{N}_0 \end{array} \right\} \le \phi(\zeta).$$

Hence from Definition 4.1.1 $u \le H_D \phi$ on D. In particular,

$$\liminf_{z \to \zeta_0} H_D \phi(z) \ge \liminf_{z \to \zeta_0} u(z) \ge \phi(\zeta_0) - \epsilon(1 + M + \phi(\zeta_0)).$$

As ϵ is arbitrary, it follows that

(4.2) $$\liminf_{z \to \zeta_0} H_D \phi(z) \ge \phi(\zeta_0).$$

Repeating the argument with ϕ replaced by $-\phi$, we also have

$$\liminf_{z \to \zeta_0} H_D(-\phi)(z) \geq -\phi(\zeta_0).$$

By Lemma 4.1.6 $H_D\phi \leq -H_D(-\phi)$, and so it follows that

(4.3) $$\limsup_{z \to \zeta_0} H_D\phi(z) \leq \phi(\zeta_0).$$

Finally, combining (4.2) and (4.3) proves the theorem. □

Putting together what we have learned, we obtain the following result.

Corollary 4.1.8 (Solution of the Dirichlet Problem) *Let D be a regular domain, and let $\phi: \partial D \to \mathbb{R}$ be a continuous function. Then there exists a unique harmonic function h on D such that $\lim_{z \to \zeta} h(z) = \phi(\zeta)$ for all $\zeta \in \partial D$.*

Proof. Uniqueness was proved in Theorem 1.2.2. For existence, take $h = H_D\phi$ and apply Theorems 4.1.2 and 4.1.5. □

There is also a converse to Theorem 4.1.5 (see the exercise below), which means that regularity is not only sufficient to guarantee solubility of the Dirichlet problem, it is also necessary. Thus Corollary 4.1.8 is, in some sense, the best possible result.

Exercises 4.1

1. Let D be a subdomain of $\mathbb{C}_\infty$ such that $\mathbb{C}_\infty \setminus D$ contains at least two points. Show that if $\zeta_0 \in \partial D$ and

$$\lim_{z \to \zeta_0} H_D\phi(z) = \phi(\zeta_0)$$

for every continuous function $\phi: \partial D \to \mathbb{R}$, then ζ_0 is a regular point. [Hint: Without loss of generality $\infty \notin \partial D$. Consider $b := H_D\phi$, where $\phi(\zeta) = -|\zeta - \zeta_0|$ $(\zeta \in \partial D)$.]

4.2 Criteria for Regularity

Although the results of the previous section appear to solve the Dirichlet problem completely, they leave one important question unanswered, namely, how to tell whether a given boundary point of D is regular? In this section we examine some geometric criteria for the existence and non-existence of barriers.

Theorem 4.2.1 *If D is a simply connected domain such that $\mathbb{C}_\infty \setminus D$ contains at least two points, then D is a regular domain.*

Proof. We need to show that every boundary point of D is regular. Given $\zeta_0 \in \partial D$, pick $\zeta_1 \in \partial D \setminus \{\zeta_0\}$. Applying a conformal mapping of the sphere, we can suppose without loss of generality that $\zeta_0 = 0$ and $\zeta_1 = \infty$. Then D is a simply connected subdomain of $\mathbb{C} \setminus \{0\}$, so by Corollary 1.1.3 there exists a holomorphic branch of $\log z$ on D. Put $N = \Delta(0,1)$, and define b on $D \cap N$ by
$$b(z) = \operatorname{Re}(1/\log z) \quad (z \in D \cap N).$$
Then b clearly satisfies all the conditions to be a barrier at 0. $\square$

This result can be 'localized' to obtain a sufficient condition for regularity of a single point.

Theorem 4.2.2 *Let D be a subdomain of $\mathbb{C}_\infty$, let $\zeta_0 \in \partial D$, and let C be the component of ∂D which contains ζ_0. If $C \neq \{\zeta_0\}$ then ζ_0 is regular.*

Proof. Choose $\zeta_1 \in C \setminus \{\zeta_0\}$. Again we can suppose that $\zeta_0 = 0$ and $\zeta_1 = \infty$. Then no closed curve in $\mathbb{C}_\infty \setminus C$ can wind around any point of C, otherwise it would disconnect C. Hence Cauchy's theorem holds in $\mathbb{C}_\infty \setminus C$, and we can repeat the proofs of Theorem 1.1.2 and Corollary 1.1.3 to obtain a holomorphic branch of $\log z$ there, and hence on D. We now proceed exactly as in the previous proof. $\square$

At the other extreme, here is a condition for irregularity.

Theorem 4.2.3 *Let D be a proper subdomain of $\mathbb{C}_\infty$, and let $\zeta_0 \in \partial D$. If there exists a neighbourhood N of ζ_0 such that $\partial D \cap N$ is polar, then ζ_0 is irregular.*

Proof. Suppose, if possible, that there exists a barrier b for ζ_0. We can assume that b is defined on $D \cap N$, where N is a connected open neighbourhood of ζ_0 such that $E := \partial D \cap \overline{N}$ is polar. Then by Theorem 3.6.3 $N \setminus E$ is still connected, and so it follows that $D \cap N = N \setminus E$. Hence by the removable singularity theorem (Theorem 3.6.1) b has a subharmonic extension to the whole of N. Since $b < 0$ on $N \setminus E$, we have $\max(b, 0) = 0$ a.e. on N, so the same equality persists everywhere, and thus $b \leq 0$ on N. Also $b(\zeta_0) \geq \limsup_{z \to \zeta_0} b(z) \geq 0$. Hence by the maximum principle $b \equiv 0$ on N, which contradicts the fact that $b < 0$ on $D \cap N$. Therefore no such barrier exists. $\square$

4.2. CRITERIA FOR REGULARITY

Theorems 4.2.2 and 4.2.3 between them provide practical tests for regularity and irregularity which cover the most commonly occurring cases. The next result, though less easy to apply, actually gives a complete characterization of regularity.

Theorem 4.2.4 *Let D be a proper subdomain of $\mathbb{C}_\infty$, and let $\zeta_0 \in \partial D$. Set $K = \mathbb{C}_\infty \setminus D$. Then the following are equivalent:*

(a) ζ_0 *is a regular boundary point of* D;

(b) K *is non-thin at* ζ_0.

If also $\infty \in D$, then these are equivalent to:

(c) K *is non-polar, and $p_\nu(\zeta_0) = I(\nu)$, where ν is the equilibrium measure for K.*

Proof. Since both the conditions (a) and (b) are invariant under conformal mapping, we can suppose from the start that $\infty \in D$, so that $K \subset \mathbb{C}$. We shall prove the implications (a) $\Rightarrow$ (b) $\Rightarrow$ (c) $\Rightarrow$ (a).

(a) $\Rightarrow$ (b): Suppose that ζ_0 is a regular point for D, with barrier b say. Let u be a function subharmonic on a neighbourhood of ζ_0, and take α with

$$\text{(4.4)} \qquad \limsup_{\substack{z \to \zeta_0 \\ z \in K \setminus \{\zeta_0\}}} u(z) < \alpha.$$

Then there exists $r > 0$ such that, if $\Delta = \Delta(\zeta_0, r)$, then u is subharmonic on a neighbourhood of $\overline{\Delta}$ and $u < \alpha$ on $\overline{\Delta} \cap K \setminus \{\zeta_0\}$. Decreasing r if necessary, we can also suppose that b is defined on a neighbourhood of $\overline{\Delta} \setminus K$. Then $\{\zeta \in \partial \Delta \setminus K : u(\zeta) \geq \alpha\}$ is a compact set on which $b < 0$, so there exists $t > 0$ such that $u + tb < \alpha$ on $\partial \Delta \setminus K$. Now for $\zeta \in \partial(\Delta \setminus K) \setminus \{\zeta_0\}$,

$$\limsup_{\substack{z \to \zeta \\ z \in \Delta \setminus K}} (u+tb)(z) \leq \begin{cases} (u+tb)(\zeta), & \zeta \in \partial \Delta \setminus K \\ u(\zeta), & \zeta \in \Delta \cap K \setminus \{\zeta_0\} \end{cases} \leq \alpha.$$

Hence by the extended maximum principle $u + tb \leq \alpha$ on $\Delta \setminus K$. Since $\lim_{z \to \zeta_0} b(z) = 0$, it follows that

$$\limsup_{\substack{z \to \zeta_0 \\ z \in \Delta \setminus K}} u(z) \leq \alpha.$$

Combining this with (4.4) leads to

$$\limsup_{\substack{z \to \zeta_0 \\ z \neq \zeta_0}} u(z) \leq \alpha.$$

Hence by the submean inequality $u(\zeta_0) \leq \alpha$. As this holds for all u, α satisfying (4.4), we conclude that K is non-thin at ζ_0.

(b) ⇒ (c): Suppose now that K is non-thin at ζ_0. From Theorem 3.8.2 it follows straightaway that K must be non-polar. Also, if ν denotes the equilibrium measure of K, then by Frostman's theorem the set

$$E := \{z \in K : p_\nu(z) > I(\nu)\}$$

is an F_σ polar set. By Theorem 3.8.2 again E is thin at ζ_0, and therefore $K \setminus E$ must be non-thin at ζ_0. Since $p_\nu = I(\nu)$ on $K \setminus E$, it follows that $p_\nu(\zeta_0) = I(\nu)$.

(c) ⇒ (a): Assume that $p_\nu(\zeta_0) = I(\nu)$. Define $b \colon D \to [-\infty, \infty)$ by

$$b(z) = I(\nu) - p_\nu(z).$$

Then b is subharmonic on D, and by Frostman's theorem $b \leq 0$ there. Since $b(\infty) = -\infty$, the maximum principle implies that in fact $b < 0$ on D. Also,

$$\liminf_{z \to \zeta_0} b(z) \geq I(\nu) - p_\nu(\zeta_0) = 0$$

by our assumption, so b is a barrier and ζ_0 is regular for D. □

This result will not be of much practical use until we have a general criterion for thinness (to be derived in Section 5.4). However, it does have some interesting theoretical consequences. The equivalence of (a) and (b), for example, explains the close correspondence between the earlier theorems in this section and the results about thinness in Section 3.8. More importantly, the equivalence of (a) and (c) shows that the set of irregular points is always small.

Theorem 4.2.5 (Kellogg's Theorem) *Let D be a proper subdomain of $\mathbb{C}_\infty$. Then the set of irregular boundary points is an F_σ polar set.*

Proof. By first performing a conformal mapping, we can suppose that $\infty \in D$. Set $K = \mathbb{C}_\infty \setminus D$. If K is polar, then by Theorem 4.2.3 every point of ∂D is irregular, and the result is clear. On the other hand, if K is non-polar, then by Theorem 4.2.4 the set of irregular points is exactly

$$\{z \in K : p_\nu(z) > I(\nu)\},$$

where ν is the equilibrium measure for K, and this is an F_σ polar set by Frostman's theorem. □

This result has a beautiful and important consequence.

4.2. CRITERIA FOR REGULARITY

Corollary 4.2.6 (Solution of the Generalized Dirichlet Problem)
Let D be a domain in $\mathbb{C}_\infty$ such that ∂D is non-polar, and let $\phi\colon \partial D \to \mathbb{R}$ be a bounded function which is continuous n.e. on ∂D. Then there exists a unique bounded harmonic function h on D such that $\lim_{z\to\zeta} h(z) = \phi(\zeta)$ for n.e. $\zeta \in \partial D$.

In order for this result to make sense, it is necessary to assume that ∂D is non-polar. However this is no great restriction, because if ∂D were polar, then by Theorem 3.6.9 (a) every bounded harmonic function on D would be constant anyway.

Proof. Set $h = H_D \phi$. Then by Theorem 4.1.2 h is harmonic and bounded on D. Also, from Theorem 4.1.5,

$$\lim_{z \to \zeta} h(z) = \phi(\zeta) \quad (\zeta \in \partial D \setminus (E_1 \cup E_2)),$$

where E_1 is the set of irregular boundary points of D, and E_2 is the set of points of discontinuity of ϕ. Now E_1 is polar by Theorem 4.2.5, and E_2 is polar by hypothesis. Also both sets are Borel, so $\lim_{z\to\zeta} h(z) = \phi(\zeta)$ for n.e. $\zeta \in \partial D$. This proves existence.

For uniqueness, suppose that h_1 and h_2 are two solutions. Then $h_1 - h_2$ is a bounded harmonic function on D satisfying

$$\lim_{z\to\zeta}(h_1 - h_2)(z) = 0 \quad \text{for n.e. } \zeta \in \partial D.$$

Applying the extended maximum principle (Theorem 3.6.9) to $\pm(h_1 - h_2)$, we deduce that $h_1 = h_2$ on D, as required. $\square$

The fact that this generalized form of the Dirichlet problem can always be solved makes it more suitable for many applications than the original form. Indeed, it will provide the basis for much of the rest of the chapter.

Exercises 4.2

1. Let D be a proper subdomain of $\mathbb{C}_\infty$, let $\zeta_0 \in \partial D \setminus \{\infty\}$, and suppose that there exists $\zeta_1 \neq \zeta_0$ such that the line segment $[\zeta_0, \zeta_1]$ does not meet D. Show that the function b, defined by

$$b(z) = -\operatorname{Re}\left[\left(\frac{z-\zeta_0}{z-\zeta_1}\right)^{1/3}\right] \quad (z \in D),$$

is a barrier at ζ_0.

4.3 Harmonic Measure

When studying the Dirichlet problem on a disc Δ in Section 1.2, we not only proved that a unique solution exists, but also gave an explicit formula for it. In the notation we have now developed, this formula may be succinctly expressed by saying that, if $\phi\colon \partial\Delta \to \mathbb{R}$ is a continuous function, then

$$H_\Delta \phi = P_\Delta \phi \quad \text{on } \Delta,$$

where $H_\Delta \phi$ and $P_\Delta \phi$ are respectively the Perron function and the Poisson integral of ϕ (see Definitions 4.1.1 and 1.2.3). We now seek to extend this to more general domains. While the Perron function is already defined for an arbitrary domain, we currently lack an appropriate analogue for the Poisson integral. To help define this, we introduce the notion of harmonic measure.

Definition 4.3.1 Let D be a proper subdomain of $\mathbb{C}_\infty$, and denote by $\mathcal{B}(\partial D)$ the σ-algebra of Borel subsets of ∂D. A *harmonic measure* for D is a function $\omega_D \colon D \times \mathcal{B}(\partial D) \to [0,1]$ such that:

(a) for each $z \in D$, the map $B \mapsto \omega_D(z, B)$ is a Borel probability measure on ∂D;

(b) if $\phi\colon \partial D \to \mathbb{R}$ is a continuous function, then $H_D\phi = P_D\phi$ on D, where $P_D\phi$ is the *generalized Poisson integral* of ϕ on D, given by

$$P_D\phi(z) := \int_{\partial D} \phi(\zeta)\, d\omega_D(z,\zeta) \quad (z \in D).$$

For example, if $\Delta = \Delta(0,1)$, then by Theorem 1.2.4

$$d\omega_\Delta(z, \zeta) := \frac{1}{2\pi} P(z, \zeta)\,|d\zeta|$$

is a harmonic measure for Δ. This reconciles the two definitions we now have for the Poisson integral $P_\Delta \phi$.

Since the definition of harmonic measure has been concocted to fit the desired conclusion, it is really only justified once we have proved the following theorem.

Theorem 4.3.2 *Let D be a domain in $\mathbb{C}_\infty$ such that ∂D is non-polar. Then there exists a unique harmonic measure ω_D for D.*

The case when ∂D is polar is less interesting—see Exercise 1 at the end of this section.

4.3. HARMONIC MEASURE

Proof. Denote by $C(\partial D)$ the space of continuous functions $\phi \colon \partial D \to \mathbb{R}$. If $\phi_1, \phi_2 \in C(\partial D)$ and $\alpha_1, \alpha_2 \in \mathbb{R}$, then $\alpha_1 H_D \phi_1 + \alpha_2 H_D \phi_2$ is a solution to the generalized Dirichlet problem with boundary data $\alpha_1 \phi_1 + \alpha_2 \phi_2$ (see Theorem 4.2.6), so by uniqueness it follows that

$$H_D(\alpha_1 \phi_1 + \alpha_2 \phi_2) = \alpha_1 H_D \phi_1 + \alpha_2 H_D \phi_2 \quad \text{on } D.$$

Also, it is clear that

$$\phi \geq 0 \text{ on } \partial D \quad \Rightarrow \quad H_D \phi \geq 0 \text{ on } D,$$
$$\phi \equiv 1 \text{ on } \partial D \quad \Rightarrow \quad H_D \phi \equiv 1 \text{ on } D.$$

Hence, for each $z \in D$, the map $\phi \mapsto H_D \phi(z)$ is a positive linear functional on $C(\partial D)$ sending the constant function 1 to 1, so by the Riesz representation theorem (Theorem A.3.2) there exists a Borel probability measure μ_z on ∂D such that

$$H_D \phi(z) = \int_{\partial D} \phi \, d\mu_z \quad (\phi \in C(\partial D)).$$

Setting

$$\omega_D(z, B) = \mu_z(B) \quad (z \in D,\ B \in \mathcal{B}(\partial D)),$$

we see immediately that (a) and (b) hold. This proves the existence of ω_D, and its uniqueness follows easily from the uniqueness part of the Riesz representation theorem. $\square$

Harmonic measure is defined so that $H_D \phi = P_D \phi$ for all continuous functions $\phi \colon \partial D \to \mathbb{R}$. The next result shows that, as a bonus, the same relation extends to a much wider class of functions ϕ.

Theorem 4.3.3 *Let D be a domain in $\mathbb{C}_\infty$ such that ∂D is non-polar. Then $H_D \phi = P_D \phi$ on D for every bounded Borel function $\phi \colon \partial D \to \mathbb{R}$.*

This gives us new information, even when D is a disc. For example, as $H_D \phi$ is always harmonic on D, the same must be true of $P_D \phi$. In the other direction, since the map $\phi \mapsto P_D \phi$ is clearly linear on bounded Borel functions, the same holds for $\phi \mapsto H_D \phi$, which was not obvious before.

Proof. First suppose that ϕ is bounded and upper semicontinuous on ∂D. Choose continuous functions $\phi_n \colon \partial D \to \mathbb{R}$ such that $\phi_n \downarrow \phi$. Then we know that $P_D \phi_n = H_D \phi_n$, so $P_D \phi_n$ is harmonic on D. From the monotone convergence theorem $P_D \phi_n \downarrow P_D \phi$ on D, and so by Harnack's theorem (Theorem 1.3.9) $P_D \phi$ is harmonic on D. Let $w \in D$ and $\epsilon > 0$. By

Definition 4.1.1, for each n we can find a subharmonic function u_n on D such that
$$\limsup_{z \to \zeta} u_n(z) \le \phi_n(\zeta) \quad (\zeta \in \partial D) \quad \text{and} \quad u_n(w) > H_D\phi_n(w) - \epsilon/2^n.$$

Define u on D by
$$u = P_D\phi + \sum_{n \ge 1}(u_n - H_D\phi_n).$$

Since $P_D\phi$ is a harmonic function and $(u_n - H_D\phi_n)$ is a negative subharmonic function for each n, it follows that u is subharmonic on D. Also, if $\zeta \in \partial D$, then for each $n \ge 1$
$$\limsup_{z \to \zeta} u(z) \le \limsup_{z \to \zeta} (P_D\phi + u_n - H_D\phi_n)(z) \le \limsup_{z \to \zeta} u_n(z) \le \phi_n(\zeta),$$
so that $\limsup_{z \to \zeta} u(z) \le \phi(\zeta)$. Hence by Definition 4.1.1 $H_D\phi \ge u$ on D. In particular,
$$H_D\phi(w) \ge u(w) \ge P_D\phi(w) - \sum_{n \ge 1} \frac{\epsilon}{2^n} = P_D\phi(w) - \epsilon.$$

As ϵ and w are arbitrary, we have shown that
$$H_D\phi \ge P_D\phi \quad \text{on } D.$$

Now suppose that ϕ is bounded and lower semicontinuous on ∂D. Applying what we have just proved to $-\phi$, we obtain
$$H_D(-\phi) \ge P_D(-\phi) \quad \text{on } D.$$

Hence, using Lemma 4.1.6, it follows that
$$H_D\phi \le -H_D(-\phi) \le -P_D(-\phi) = P_D\phi \quad \text{on } D.$$

Finally, suppose that ϕ is an arbitrary bounded Borel function on ∂D. Let $w \in D$ and $\epsilon > 0$. Then, as the Borel probability measure $\omega_D(w, \cdot)$ is regular, we can appeal to the Vitali–Carathéodory theorem (Theorem A.2.3 in the Appendix) to obtain an upper semicontinuous function ψ_u and a lower semicontinuous function ψ_l on ∂D such that
$$\psi_u \le \phi \le \psi_l \quad \text{and} \quad \int_{\partial D} (\psi_l - \psi_u)(\zeta)\, d\omega_D(w, \zeta) < \epsilon.$$

Replacing ψ_u by $\max(\psi_u, -\|\phi\|_\infty)$ and ψ_l by $\min(\psi_l, \|\phi\|_\infty)$, we can further suppose that ψ_u and ψ_l are bounded on ∂D. Then by what we have already proved,
$$H_D\psi_u \ge P_D\psi_u \quad \text{and} \quad H_D\psi_l \le P_D\psi_l \quad \text{on } D.$$

4.3. HARMONIC MEASURE

Therefore

$$H_D\phi(w) \leq H_D\psi_l(w) \leq P_D\psi_l(w) \leq P_D\psi_u(w) + \epsilon \leq P_D\phi(w) + \epsilon$$

and

$$H_D\phi(w) \geq H_D\psi_u(w) \geq P_D\psi_u(w) \geq P_D\psi_l(w) - \epsilon \geq P_D\phi(w) - \epsilon.$$

As ϵ and w are arbitrary, we conclude that $H_D\phi = P_D\phi$ on D. □

From this result we can deduce a characterization of harmonic measure which explains its nomenclature.

Theorem 4.3.4 *Let D be a domain in $\mathbb{C}_\infty$ such that ∂D is non-polar, and let B be a Borel subset of ∂D. Then:*

(a) *the function $z \mapsto \omega_D(z, B)$ is harmonic and bounded on D;*

(b) *if ζ is a regular boundary point of D which lies outside the relative boundary of B in ∂D, then*

$$\lim_{z \to \zeta} \omega_D(z, B) = 1_B(\zeta).$$

Moreover, if the relative boundary of B in ∂D is polar, then the function $\omega_D(\cdot, B)$ is uniquely determined by (a) and (b).

Proof. By Theorem 4.3.3 we have

$$\omega_D(z, B) = H_D 1_B(z) \quad (z \in D).$$

Therefore (a) follows immediately from Theorem 4.1.2. Also if ζ satisfies the hypotheses of (b), then 1_B is continuous at ζ, and so the conclusion of (b) follows from Theorem 4.1.5.

The uniqueness part of the result is an immediate consequence of Corollary 4.2.6. □

This theorem shows that, provided the relative boundary of B in ∂D is polar, the function $\omega_D(\cdot, B)$ is exactly the solution of the generalized Dirichlet problem with boundary data $\phi = 1_B$. This provides a quick way of identifying the harmonic measure in a number of important special cases—one simply 'spots' a harmonic function with the right boundary values. Some examples are given in Table 4.1.

Table 4.1: Examples of Harmonic Measure

D	B	$\omega_D(z, B)$
$\{\operatorname{Im} z > 0\}$	$[a, b]$	$\dfrac{1}{\pi} \arg\left(\dfrac{z-b}{z-a}\right)$
$\{|z| < 1, \operatorname{Im} z > 0\}$	$\{|z| = 1, \operatorname{Im} z > 0\}$	$\dfrac{2}{\pi} \arg\left(\dfrac{1+z}{1-z}\right)$
$\{a < \operatorname{Re} z < b\}$	$\{\operatorname{Re} z = b\}$	$\dfrac{\operatorname{Re} z - a}{b - a}$
$\{\alpha < \arg z < \beta\}$	$\{\arg z = \beta\}$	$\dfrac{\arg z - \alpha}{\beta - \alpha}$
$\{r < |z| < s\}$	$\{|z| = s\}$	$\dfrac{\log(|z|/r)}{\log(s/r)}$

Theorem 4.3.4 also has another interesting consequence.

Corollary 4.3.5 *Let D be a domain in $\mathbb{C}_\infty$ such that ∂D is non-polar. Then the measures $\{\omega_D(z, \cdot) : z \in D\}$ are mutually absolutely continuous. In fact, if $z, w \in D$, then*

$$\omega_D(z, B) \leq \tau_D(z, w)\omega_D(w, B) \quad (B \in \mathcal{B}(\partial D)),$$

where $\tau_D(z, w)$ is the Harnack distance between z and w.

Proof. We recall from Definition 1.3.4 that $h(z) \leq \tau_D(z, w)h(w)$ for every positive harmonic function h on D. The result follows by applying this with $h = \omega_D(\cdot, B)$. □

It thus makes sense to describe subsets of ∂D as having harmonic measure zero without referring to a particular base point $z \in D$. The next result gives some examples of these.

Theorem 4.3.6 *Let D be a domain in $\mathbb{C}_\infty$ such that ∂D is non-polar. Then every Borel polar subset of ∂D has harmonic measure zero.*

4.3. HARMONIC MEASURE

Proof. Let E be a Borel polar subset of ∂D. If u is a subharmonic function on D such that $\limsup_{z\to\zeta} u(z) \le 1_E(\zeta)$ ($\zeta \in \partial D$), then by the extended maximum principle $u \le 0$ on D. It follows that $H_D 1_E \equiv 0$ on D, and hence by Theorem 4.3.3 $P_D 1_E \equiv 0$ on D. □

It is reasonable to ask whether, conversely, every set of harmonic measure zero for D must be polar. The answer is no, though this will only become apparent later (see Exercise 5.3.7).

We now prove two basic general inequalities involving harmonic measure, one for subharmonic functions, and one for meromorphic ones. In view of Theorem 4.3.6, the first of these is a generalization of the extended maximum principle.

Theorem 4.3.7 (Two-Constant Theorem) *Let D be a domain in $\mathbb{C}_\infty$ such that ∂D is non-polar, and let B be a Borel subset of ∂D. If u is subharmonic on D and satisfies*

$$u(z) \le M \quad (z \in D) \quad \text{and} \quad \limsup_{z\to\zeta} u(z) \le m \quad (\zeta \in B),$$

where M and m are constants, then

$$u(z) \le m\omega_D(z, B) + M(1 - \omega_D(z, B)) \quad (z \in D).$$

Proof. Set $\phi = m 1_B + M(1 - 1_B)$ on ∂D. Then $\limsup_{z\to\zeta} u(z) \le \phi(\zeta)$ for all $\zeta \in \partial D$, so by Definition 4.1.1 $u \le H_D \phi$ on D. From Theorem 4.3.3 it follows that

$$u \le P_D \phi = m P_D 1_B + M(1 - P_D 1_B) \quad \text{on } D,$$

which gives the desired inequality. □

Theorem 4.3.8 (Subordination Principle) *Let D_1 and D_2 be domains in $\mathbb{C}_\infty$ with non-polar boundaries, and let B_1 and B_2 be Borel subsets of ∂D_1 and ∂D_2 respectively. Let $f: D_1 \cup B_1 \to D_2 \cup B_2$ be a continuous map which is meromorphic on D_1, and suppose that $f(D_1) \subset D_2$ and $f(B_1) \subset B_2$. Then*

$$\omega_{D_2}(f(z), B_2) \ge \omega_{D_1}(z, B_1) \quad (z \in D_1),$$

with equality if f is also a homeomorphism of $D_1 \cup B_1$ onto $D_2 \cup B_2$.

Proof. Set $\phi_1 = 1 - 1_{B_1}$ and $\phi_2 = 1 - 1_{B_2}$ on ∂D_1 and ∂D_2 respectively. Let u be a subharmonic function on D_2 such that

$$\limsup_{z\to\zeta} u(z) \le \phi_2(\zeta) \quad (\zeta \in \partial D_2).$$

Then it is readily checked that $u \circ f$ is a subharmonic function on D_1 such that
$$\limsup_{z \to \zeta}(u \circ f)(z) \le \phi_1(\zeta) \quad (\zeta \in \partial D_1),$$
and therefore $u \circ f \le H_{D_1}\phi_1$ on D_1. As this holds for all such u, we deduce that
$$(H_{D_2}\phi_2) \circ f \le H_{D_1}\phi_1 \quad \text{on } D_1.$$
By Theorem 4.3.3 $H_{D_j}\phi_j = P_{D_j}\phi_j = 1 - P_{D_j}1_{B_j}$ $(j=1,2)$, and hence
$$(P_{D_2}1_{B_2}) \circ f \ge P_{D_1}1_{B_1} \quad \text{on } D_1,$$
which gives the desired inequality.

If f is also a homeomorphism of $D_1 \cup B_1$ onto $D_2 \cup B_2$, then we can apply the same argument to f^{-1} to deduce that equality holds. $\square$

Corollary 4.3.9 *Let D_1 and D_2 be domains in $\mathbb{C}_\infty$ with non-polar boundaries, and suppose that $D_1 \subset D_2$. If B is a Borel subset of $\partial D_1 \cap \partial D_2$, then*
$$\omega_{D_1}(z, B) \le \omega_{D_2}(z, B) \quad (z \in D_1).$$

Proof. Take $f: D_1 \cup B \to D_2 \cup B$ to be the inclusion map. $\square$

As an application of these ideas, we shall prove a theorem about asymptotic values.

Definition 4.3.10 *Let ϕ be a function defined on an unbounded domain D in $\mathbb{C}$. Then a is an asymptotic value of ϕ if there exists a path $\Gamma: [0, \infty) \to D$ such that*
$$\lim_{t \to \infty} \Gamma(t) = \infty \quad \text{and} \quad \lim_{t \to \infty} \phi(\Gamma(t)) = a.$$

Theorem 4.3.11 *Let u be a subharmonic function on $H := \{z : \operatorname{Im} z > 0\}$ such that $u < 0$ on H. If $a \in [-\infty, 0)$ is an asymptotic value of u, then for each $\alpha \in (0, \pi/2]$*
$$\limsup_{\substack{z \to \infty \\ z \in S_\alpha}} u(z) \le \frac{\alpha}{\pi}a,$$
where S_α is the sector $\{z \in H : \alpha \le \arg z \le \pi - \alpha\}$.

Proof. Let $\Gamma: [0, \infty) \to H$ be a path such that
$$\lim_{t \to \infty} \Gamma(t) = \infty \quad \text{and} \quad \lim_{t \to \infty} u(\Gamma(t)) = a.$$
Take a' with $a < a' < 0$, and choose $R > 0$ so large that $u < a'$ on $\Gamma \cap D_R$, where $D_R = \{z \in H : |z| > R\}$. We may also suppose that Γ meets the

4.3. HARMONIC MEASURE

circle $\{|z| = R\}$. Fix $z \in D_R \setminus \Gamma$, and let W be the component of $D_R \setminus \Gamma$ containing z. Then since $u \leq a'$ on $\partial W \setminus \partial D_R$, the two-constant theorem (Theorem 4.3.7) gives

$$u(z) \leq a' \omega_W(z, \partial W \setminus \partial D_R).$$

We now seek to estimate the right-hand side of this inequality. Notice that since $a' < 0$, this means finding a *lower* bound for the harmonic measure. Using Corollary 4.3.9, we have

$$\begin{aligned} \omega_W(z, \partial W \setminus \partial D_R) &= 1 - \omega_W(z, \partial W \cap \partial D_R) \\ &\geq 1 - \omega_{D_R}(z, \partial W \cap \partial D_R) \\ &= \omega_{D_R}(z, \partial D_R \setminus \partial W). \end{aligned}$$

Now ∂W cannot meet both $(-\infty, -R]$ and $[R, \infty)$, for then Γ would disconnect the connected set W. If $\partial W \cap (-\infty, -R] = \emptyset$, then

$$\begin{aligned} \omega_{D_R}(z, \partial D_R \setminus \partial W) &\geq \omega_{D_R}(z, (-\infty, -R]) \\ &= H_{D_R} 1_{(-\infty, -R]}(z) \\ &\geq \frac{1}{\pi} \arg(z) - H_{D_R} 1_{C_R}(z), \end{aligned}$$

where $C_R = \{\zeta \in \partial D_R : |\zeta| = R\}$. If, on the other hand, $\partial W \cap (R, \infty] = \emptyset$, then a similar argument shows that

$$\omega_{D_R}(z, \partial D_R \setminus \partial W) \geq \frac{1}{\pi}(\pi - \arg(z)) - H_{D_R} 1_{C_R}(z).$$

Combining all these estimates, we finally arrive at the conclusion that

$$u(z) \leq a' \frac{1}{\pi} \min(\arg z, \pi - \arg z) - a' H_{D_R} 1_{C_R}(z).$$

Note that, although this inequality was proved under the assumption that $z \in D_R \setminus \Gamma$, it evidently holds if $z \in D_R \cap \Gamma$ as well. Hence, in particular,

$$u(z) \leq a' \frac{\alpha}{\pi} - a' H_{D_R} 1_{C_R}(z) \quad (z \in D_R \cap S_\alpha).$$

Now $1_{C_R} = 0$ on a neighbourhood of ∞, which is a regular boundary point of D_R, so by Theorem 4.1.5 $\lim_{z \to \infty} H_{D_R} 1_{C_R}(z) = 0$. We therefore deduce that

$$\limsup_{\substack{z \to \infty \\ z \in S_\alpha}} u(z) \leq a' \frac{\alpha}{\pi},$$

and since this holds for each $a' > a$, the desired result follows. $\square$

The harmonic function $u = -\arg z$, which has $-\pi$ as an asymptotic value, shows that the above theorem is sharp. Of course, this function also has many other asymptotic values. By contrast, a bounded holomorphic function on H can have at most one.

Corollary 4.3.12 (Lindelöf's Theorem) *Let f be a bounded holomorphic function on $H := \{z : \operatorname{Im} z > 0\}$. If a is an asymptotic value of f then, for each sector S_α as in Theorem 4.3.11, $f(z) \to a$ uniformly as $z \to \infty$ in S_α. In particular, f can have at most one asymptotic value.*

Proof. Apply Theorem 4.3.11 with $u = \log(|f-a|/M)$, where $M = \sup_H |f-a|$. □

These results provide a good illustration of how many problems in potential theory and complex analysis can be reduced to questions about harmonic measure. It is therefore of great importance to be able to compute, or at least estimate, harmonic measure for as many domains as possible. Simple cases can be treated using conformal mapping. As an illustration, we now compute the important example of harmonic measure for the half-plane.

Theorem 4.3.13 *Let $H = \{z \in \mathbb{C} : \operatorname{Im} z > 0\}$. If B is a Borel subset of $\mathbb{R}$, then*

$$\omega_H(x+iy, B) = \frac{1}{\pi} \int_B \frac{y\, dt}{(x-t)^2 + y^2} \quad (x+iy \in H).$$

Proof. Set $\Delta = \Delta(0,1)$, and let $f : H \to \Delta$ be the conformal mapping

$$f(z) = \frac{z-i}{z+i} \quad (z \in H).$$

Then by Theorem 4.3.8,

$$\begin{aligned}
\omega_H(z,B) &= \omega_\Delta(f(z), f(B)) \\
&= \frac{1}{2\pi} \int_{f(B)} \frac{1 - |f(z)|^2}{|\zeta - f(z)|^2} |d\zeta| \\
&= \frac{1}{2\pi} \int_B \frac{1 - |f(z)|^2}{|f(t) - f(z)|^2} |f'(t)|\, dt \\
&= \frac{1}{\pi} \int_B \frac{\operatorname{Im} z}{|z-t|^2}\, dt.
\end{aligned}$$

Substituting $z = x+iy$ gives the result. □

4.3. HARMONIC MEASURE

The problem of estimating harmonic measure for more complicated domains is a vast subject, well beyond the scope of a book such as this. We shall content ourselves with one general estimate for simply connected domains, which will be obtained later as a by-product of the solution of the Carleman–Milloux problem (see Corollary 4.5.9).

We conclude this section by relating harmonic measure to equilibrium measure.

Theorem 4.3.14 *Let K be a compact non-polar subset of $\mathbb{C}$. Then its equilibrium measure ν is given by*

$$\nu = \omega_D(\infty, \cdot),$$

where D is the component of $\mathbb{C}_\infty \setminus K$ containing ∞.

Proof. Write ω for $\omega_D(\infty, \cdot)$, so ω is a Borel probability measure on K. If we define

$$u(z) = \begin{cases} p_\omega(z) - p_\nu(z) + I(\nu), & z \in D \setminus \{\infty\}, \\ I(\nu), & z = \infty, \end{cases}$$

then u is subharmonic on D, and $\limsup_{z \to \zeta} u(z) \leq p_\omega(\zeta)$ for all $\zeta \in \partial D$. Writing $\phi_n := \max(p_\omega, -n)$ on ∂D, it follows that $u \leq H_D \phi_n = P_D \phi_n$ on D, and letting $n \to \infty$ we deduce that

$$u(z) \leq \int_{\partial D} p_\omega(\zeta) \, d\omega_D(z, \zeta) \quad (z \in D).$$

In particular, putting $z = \infty$, we obtain $I(\nu) \leq I(\omega)$. This implies that ω is an equilibrium measure for K, and so by uniqueness (Theorem 3.7.6) it follows that $\omega = \nu$. □

Exercises 4.3

1. Let D be a domain in $\mathbb{C}_\infty$ such that ∂D is polar. Show that, if ∂D contains at least two points, then the map $\phi \mapsto H_D \phi$ is not linear, and deduce that no harmonic measure for D exists. What happens if ∂D contains just one point?

2. Let D be a domain in $\mathbb{C}_\infty$ such that ∂D is non-polar. Show that if $\phi: \partial D \to \mathbb{R}$ is a bounded Borel function then

$$\liminf_{z \to \zeta} H_D \phi(z) \leq \phi(\zeta) \leq \limsup_{z \to \zeta} H_D \phi(z)$$

for all $\zeta \in \partial D$ outside a set of harmonic measure zero.
[Hint: put $\psi(\zeta) = \min(\phi(\zeta), \limsup_{z \to \zeta} H_D \phi(z))$ ($\zeta \in \partial D$), and show that $H_D \phi \leq H_D \psi$.]

3. Show that $f(z) = \exp(i\log(z/i))$ is a bounded holomorphic function on $\{z : \operatorname{Im} z > 0\}$, but that it has no asymptotic values there.

4. Re-derive the formula for harmonic measure for the half-plane (Theorem 4.3.13) using the appropriate entry in Table 4.1.

4.4 Green's Functions

The harmonic measure of a domain is intimately related to another important invariant, the Green's function. In essence, a Green's function is a family of fundamental solutions of the Laplacian, each of which is zero on the boundary. Here is the precise definition.

Definition 4.4.1 Let D be a proper subdomain of $\mathbb{C}_\infty$. A *Green's function* for D is a map $g_D \colon D \times D \to (-\infty, \infty]$, such that for each $w \in D$:

(a) $g_D(\cdot, w)$ is harmonic on $D \setminus \{w\}$, and bounded outside each neighbourhood of w;

(b) $g_D(w, w) = \infty$, and as $z \to w$,
$$g_D(z, w) = \begin{cases} \log|z| + O(1), & w = \infty, \\ -\log|z - w| + O(1), & w \neq \infty; \end{cases}$$

(c) $g_D(z, w) \to 0$ as $z \to \zeta$, for n.e. $\zeta \in \partial D$.

For example, if $\Delta = \Delta(0, 1)$, then
$$g_\Delta(z, w) := \log\left|\frac{1 - z\overline{w}}{z - w}\right|$$
is a Green's function for Δ.

As usual, to justify the definition we need an existence-and-uniqueness theorem.

Theorem 4.4.2 *If D is a domain in $\mathbb{C}_\infty$ such that ∂D is non-polar, then there exists a unique Green's function g_D for D.*

Once again, the case when ∂D is polar is less interesting—see Exercise 1 at the end of the section.

Proof. We begin with uniqueness. Suppose that g_1 and g_2 are two Green's functions for D. Given $w \in D$, define
$$h(z) = g_1(z, w) - g_2(z, w) \quad (z \in D \setminus \{w\}).$$

4.4. GREEN'S FUNCTIONS

Then h is harmonic and bounded on $D \setminus \{w\}$, and $\lim_{z \to \zeta} h(z) = 0$ for n.e. $\zeta \in \partial D$, so by the extended maximum principle (Theorem 3.6.9) $h \equiv 0$ on $D \setminus \{w\}$. As this holds for each $w \in D$, we deduce that $g_1 = g_2$ on $D \times D$.

We first prove the existence of $g_D(z, w)$ when $w = \infty \in D$. Set $K = \mathbb{C}_\infty \setminus D$, so that K is a compact non-polar subset of $\mathbb{C}$, and let ν be its equilibrium measure. If we define

$$g_D(z, \infty) = \begin{cases} p_\nu(z) - I(\nu), & z \in D \setminus \{\infty\}, \\ \infty, & z = \infty, \end{cases}$$

then, with the help of Frostman's theorem, it is easily checked that $g_D(\cdot, \infty)$ satisfies the conditions (a), (b), (c) of Definition 4.4.1 with $w = \infty$.

Now for $w \in D$, $w \neq \infty$, define

$$g_D(z, w) = g_{D'}\left(\frac{1}{z - w}, \infty\right) \quad (z \in D),$$

where D' is the image of D under the map $z \mapsto (z - w)^{-1}$. Applying what we have already proved to the domain D', it follows that $g_D(\cdot, w)$ satisfies (a), (b), (c) of Definition 4.4.1. Hence $g_D(z, w)$ exists for all $z, w \in D$. $\square$

We now start to investigate the properties of Green's functions. The most basic one is positivity.

Theorem 4.4.3 *Let D be a domain in $\mathbb{C}_\infty$ such that ∂D is non-polar. Then*

$$g_D(z, w) > 0 \quad (z, w \in D).$$

Proof. Fix $w \in D$, and define

$$u(z) = -g_D(z, w) \quad (z \in D).$$

Then u is subharmonic and bounded above on D, and $\limsup_{z \to \zeta} u(z) = 0$ for n.e. $\zeta \in \partial D$. Hence by the extended maximum principle $u \leq 0$ on D. Moreover, if it were the case that $u(z) = 0$ for some $z \in D$, then by the standard maximum principle it would follow that $u \equiv 0$ on D, which is false: for example $u(w) = -g_D(w, w) = -\infty$. Hence $u < 0$ on D, which proves the result. $\square$

As with the Harnack distance and harmonic measure, Green's functions admit a subordination principle for meromorphic functions.

Theorem 4.4.4 (Subordination Principle) *Let D_1 and D_2 be domains in $\mathbb{C}_\infty$ with non-polar boundaries, and let $f: D_1 \to D_2$ be a meromorphic function. Then*

$$g_{D_2}(f(z), f(w)) \geq g_{D_1}(z, w) \quad (z, w \in D_1),$$

with equality if f is a conformal mapping of D_1 onto D_2.

Proof. Assume first that $w \neq \infty$ and $f(w) \neq \infty$, and define

$$u(z) = g_{D_1}(z,w) - g_{D_2}(f(z), f(w)) \quad (z \in D_1 \setminus \{w\}).$$

Then u is subharmonic on $D_1 \setminus \{w\}$. Also u is bounded above outside each neighbourhood of w, and as $z \to w$,

$$u(z) = \log\left|\frac{f(z) - f(w)}{z - w}\right| + O(1) = \log|f'(w)| + O(1),$$

so in fact u is bounded above on $D_1 \setminus \{w\}$. Lastly, since $g_{D_2} > 0$,

$$\limsup_{z \to \zeta} u(z) \leq \lim_{z \to \zeta} g_{D_1}(z,w) = 0 \quad \text{for n.e. } \zeta \in \partial D_1.$$

Hence by the extended maximum principle $u \leq 0$ on $D_1 \setminus \{w\}$, which gives the desired inequality.

For the case when $w = \infty$ and $f(w) \neq \infty$, we recall from the construction of the Green's function that

$$g_{D_1}(z, \infty) = g_{D_1'}(1/z, 0),$$

where D_1' is the image of D_1 under the map $z \mapsto 1/z$. Hence the result follows by applying the previous case to the function $z \mapsto f(1/z) \colon D_1' \to D_2$. The case when $f(w) = \infty$ is treated similarly.

Finally, if f is a conformal mapping of D_1 onto D_2, then we can apply the inequality already proved to f^{-1} to deduce that equality holds. $\square$

This result allows us to compute Green's functions for some simple domains by means of conformal mapping. A few examples are given in Table 4.2.

Another consequence of Theorem 4.4.4 is that g_D increases with D.

Corollary 4.4.5 *Let D_1 and D_2 be domains in $\mathbb{C}_\infty$ with non-polar boundaries. If $D_1 \subset D_2$ then*

$$g_{D_1}(z,w) \leq g_{D_2}(z,w) \quad (z, w \in D_1).$$

Proof. Take $f \colon D_1 \to D_2$ to be the inclusion map. $\square$

In fact g_D increases continuously with D, in the following sense.

Theorem 4.4.6 *Let D be a domain in $\mathbb{C}_\infty$ such that ∂D is non-polar, and let $(D_n)_{n \geq 1}$ be subdomains of D such that $D_1 \subset D_2 \subset D_3 \subset \cdots$ and $\cup_n D_n = D$. Then*

$$\lim_{n \to \infty} g_{D_n}(z,w) = g_D(z,w) \quad (z, w \in D).$$

4.4. GREEN'S FUNCTIONS

Table 4.2: Examples of Green's Functions

D	$g_D(z,w)$
$\{|z|<\rho\}$	$\log\left|\dfrac{\rho^2 - z\overline{w}}{\rho(z-w)}\right|$
$\{\operatorname{Im} z > 0\}$	$\log\left|\dfrac{z-\overline{w}}{z-w}\right|$
$\{\operatorname{Re} z > 0\}$	$\log\left|\dfrac{z+\overline{w}}{z-w}\right|$
$\{|\arg z| < \pi/(2\alpha)\}$	$\log\left|\dfrac{z^\alpha + \overline{w}^\alpha}{z^\alpha - w^\alpha}\right|$
$\{|\operatorname{Re} z| < \pi/(2\alpha)\}$	$\log\left|\dfrac{e^{i\alpha z} + e^{-i\alpha \overline{w}}}{e^{i\alpha z} - e^{i\alpha w}}\right|$

Proof. Fix $w \in D$. Then $w \in D_{n_0}$ for some n_0, and by renumbering the sequence (D_n), we may suppose that $n_0 = 1$. For $n \geq 1$ define
$$h_n(z) = g_D(z,w) - g_{D_n}(z,w) \quad (z \in D_n \setminus \{w\}).$$
Then h_n is harmonic on $D_n \setminus \{w\}$ and bounded near w, so by the removable singularity theorem (Corollary 3.6.2) h_n extends to be harmonic on D_n. Corollary 4.4.5 implies that $h_n \geq h_{n+1}$ on D_n for each n, so $u := \lim_{n \to \infty} h_n$ is subharmonic on D. Since $h_n \leq g_D(\cdot,w)$ on D_n for each n, it follows that $u \leq g_D(\cdot,w)$ on D. Hence u is bounded above on D, and also $\limsup_{z \to \zeta} u(z) \leq 0$ for n.e. $\zeta \in \partial D$. Therefore by the extended maximum principle $u \leq 0$ on D. This tells us that
$$\liminf_{n \to \infty} g_{D_n}(z,w) \geq g_D(z,w) \quad (z \in D).$$
But from Corollary 4.4.5 we also have
$$\limsup_{n \to \infty} g_{D_n}(z,w) \leq g_D(z,w) \quad (z \in D).$$
Combining these two inequalities yields the result. $\square$

For bounded domains there is an integral formula for the Green's function in terms of the harmonic measure.

Theorem 4.4.7 *Let D be a bounded domain in $\mathbb{C}$. Then*
$$g_D(z,w) = \int_{\partial D} \log|\zeta - w|\, d\omega_D(z,\zeta) - \log|z-w| \quad (z, w \in D).$$

Proof. Given $w \in D$, define $\phi_w : \partial D \to \mathbb{R}$ by
$$\phi_w(\zeta) = \log|\zeta - w| \quad (\zeta \in \partial D).$$
Then $P_D \phi_w$ is harmonic and bounded on D, and $\lim_{z \to \zeta} P_D \phi_w(z) = \phi_w(\zeta)$ for n.e. $\zeta \in \partial D$. Therefore the function
$$(z,w) \mapsto P_D \phi_w(z) - \log|z-w|$$
satisfies conditions (a), (b), (c) of Definition 4.4.1, and so by uniqueness it must be the Green's function g_D. □

The importance of this formula is that it tells us how $g_D(z,w)$ depends on w, which is the key to proving the following symmetry theorem for Green's functions. In view of the asymmetric way that $g_D(z,w)$ was defined, this is perhaps a surprising result.

Theorem 4.4.8 (Symmetry Theorem) *Let D be a domain in $\mathbb{C}_\infty$ such that ∂D is non-polar. Then*
$$g_D(z,w) = g_D(w,z) \quad (z,w \in D).$$

Proof. Applying a conformal mapping, we can suppose that $D \subset \mathbb{C}$. Then D can be exhausted by an increasing sequence of bounded subdomains (D_n), and by Theorem 4.4.6, g_D will be symmetric provided that each g_{D_n} is symmetric. It is thus sufficient to prove the result in the case when D is a bounded subdomain of $\mathbb{C}$.

Fix such a domain D, and let $w \in D$. Define u on $D \setminus \{w\}$ by
$$u(z) = g_D(z,w) - g_D(w,z) \quad (z \in D \setminus \{w\}).$$
Switching the rôles of z and w in Theorem 4.4.7, we have
$$u(z) = g_D(z,w) + \log|z - w| - \int_{\partial D} \log|\zeta - z|\, d\omega_D(w,\zeta) \quad (z \in D \setminus \{w\}).$$
With the help of Theorem 2.4.8, this formula shows that u is subharmonic on $D \setminus \{w\}$. It also tell us that u is bounded above there. In addition, from the original definition of u, we have
$$\limsup_{z \to \zeta} u(z) \le \lim_{z \to \zeta} g_D(z,w) = 0 \quad \text{for n.e. } \zeta \in \partial D.$$

4.4. GREEN'S FUNCTIONS

Hence by the extended maximum principle $u \leq 0$ on $D \setminus \{w\}$. Thus

$$g_D(z,w) \leq g_D(w,z) \quad (z \in D),$$

and since w is arbitrary, the result follows. □

As part of the definition of Green's function, $\lim_{z \to \zeta} g_D(z,w) = 0$ for n.e. $\zeta \in \partial D$, but it is not clear whether the exceptional set depends on w. The next result shows that it doesn't, and identifies it precisely.

Theorem 4.4.9 *Let D be a domain in $\mathbb{C}_\infty$ such that ∂D is non-polar, let $w \in D$, and let $\zeta \in \partial D$. Then*

$$\lim_{z \to \zeta} g_D(z,w) = 0$$

if and only if ζ is a regular boundary point of D.

Proof. If $\lim_{z \to \zeta} g_D(z,w) = 0$, then $-g_D(\cdot, w)$ is a barrier at ζ, and so ζ is regular.

Conversely, suppose that ζ is a regular boundary point of D. Let N be a relatively compact neighbourhood of w in D, and define $\phi: \partial(D \setminus \overline{N}) \to \mathbb{R}$ by

$$\phi(\zeta) = \begin{cases} 0, & \zeta \in \partial D, \\ g_D(\zeta, w), & \zeta \in \partial N. \end{cases}$$

Then $g_D(\cdot, w)$ solves the generalized Dirichlet problem on $D \setminus \overline{N}$ with boundary function ϕ (see Corollary 4.2.6), so by uniqueness it follows that

$$g_D(z,w) = H_{D \setminus \overline{N}} \phi(z) \quad (z \in D \setminus \overline{N}).$$

Hence, as ζ is a regular point for D, and thus also for $D \setminus \overline{N}$, Theorem 4.1.5 implies that $\lim_{z \to \zeta} g_D(z,w) = \phi(\zeta) = 0$. □

This result provides a characterization of regular points which is internal to D. This has some interesting consequences—see for example Exercise 6 below. We now use it, together with the symmetry property of Green's functions, to prove a strong converse to the subordination principle stated earlier.

Theorem 4.4.10 *Let D_1 and D_2 be domains in $\mathbb{C}_\infty$ with non-polar boundaries, and let $f\colon D_1 \to D_2$ be a meromorphic function.*

(a) *If there exist distinct points $z_0, w_0 \in D_1$ such that*
$$g_{D_2}(f(z_0), f(w_0)) = g_{D_1}(z_0, w_0),$$
then
$$g_{D_2}(f(z), f(w)) = g_{D_1}(z, w)$$
for all $z, w \in D_1$, and f is injective.

(b) *If, further, D_1 is a regular domain, then f is also surjective, and is therefore a conformal mapping of D_1 onto D_2.*

Proof. (a) Define
$$u(z) = g_{D_1}(z, w_0) - g_{D_2}(f(z), f(w_0)) \quad (z \in D_1 \setminus \{w_0\}).$$
Then u is subharmonic on $D_1 \setminus \{w_0\}$, and by Theorem 4.4.4 $u \le 0$ there. Since, by hypothesis, $u(z_0) = 0$, it follows from the maximum principle that $u \equiv 0$, and hence
$$g_{D_2}(f(z), f(w_0)) = g_{D_1}(z, w_0) \quad (z \in D_1).$$
We can now use Theorem 4.4.8 to switch the rôles of z and w, and repeat the argument to obtain
$$g_{D_2}(f(z), f(w)) = g_{D_1}(z, w) \quad (z, w \in D_1).$$
This implies that f is injective, since
$$z \ne w \Rightarrow g_{D_1}(z, w) < \infty \Rightarrow g_{D_2}(f(z), f(w)) < \infty \Rightarrow f(z) \ne f(w).$$

(b) Suppose that $f(D_1) \ne D_2$. Then an elementary connectedness argument shows that $\partial f(D_1) \cap D_2 \ne \emptyset$. Let η be a point in this set, and choose (z_n) in D_1 such that $f(z_n) \to \eta$. Replacing (z_n) by a subsequence, if necessary, we may also suppose that $z_n \to \zeta \in \partial D_1$. Then for any $w \in D_1$, we have
$$\lim_{n \to \infty} g_{D_1}(z_n, w) = \lim_{n \to \infty} g_{D_2}(f(z_n), f(w)) = g_{D_2}(\eta, f(w)) > 0,$$
so by Theorem 4.4.9 ζ must be an irregular point of ∂D_1. Thus if D_1 is a regular domain, then necessarily $f(D_1) = D_2$. $\square$

4.4. GREEN'S FUNCTIONS

As an application of this result, we obtain a simple proof of the Riemann mapping theorem.

Theorem 4.4.11 (Riemann Mapping Theorem) *If D is a simply connected proper subdomain of $\mathbb{C}$, then there exists a conformal map of D onto the unit disc Δ.*

Proof. By Theorem 4.2.1 D is a regular domain. In particular, ∂D is non-polar, so D has a Green's function g_D. Fix $w \in D$, and define

$$h(z) = g_D(z, w) + \log |z - w| \quad (z \in D \setminus \{w\}).$$

Then h is harmonic on $D \setminus \{w\}$ and bounded near w, so by the removable singularity theorem (Corollary 3.6.2) h extends to be harmonic on D. Applying Theorem 1.1.2, we can write $h = \operatorname{Re} f_1$, for some holomorphic function f_1 on D. Define

$$f(z) = (z - w)e^{-f_1(z)} \quad (z \in D).$$

Then f is holomorphic on D and $f(w) = 0$. Also

$$\log |f(z)| = -g_D(z, w) \quad (z \in D),$$

which shows that f maps D into Δ, and that

$$g_\Delta(f(z), f(w)) = g_D(z, w) \quad (z \in D).$$

Theorem 4.4.10 can now be applied to deduce that f is actually a conformal mapping of D onto Δ. $\square$

In general, the conformal map $f: D \to \Delta$ will not extend to a homeomorphism of the closures. For this to be possible, it is clear that every boundary point of D must be accessible, in the following sense.

Definition 4.4.12 A point $\zeta \in \partial D$ is *accessible* if, for each sequence (z_n) in D with $\lim_{n \to \infty} z_n = \zeta$, there exists a path $\Gamma: [0, \infty) \to D$ with $\lim_{t \to \infty} \Gamma(t) = \zeta$, such that $\Gamma(t_n) = z_n$ for some increasing sequence $t_n \to \infty$.

It turns out that this simple necessary condition is also sufficient.

Theorem 4.4.13 *Let D be a bounded simply connected domain in $\mathbb{C}$, and let $f: D \to \Delta$ be a conformal mapping of D onto the unit disc Δ.*

(a) *If $\zeta \in \partial D$ is accessible, then f extends continuously to $D \cup \{\zeta\}$, and $|f(\zeta)| = 1$.*

(b) *If $\zeta, \zeta' \in \partial D$ are distinct accessible points, then $f(\zeta) \neq f(\zeta')$.*

(c) *If every boundary point of D is accessible, then f extends to a homeomorphism of $\overline{D}$ onto $\overline{\Delta}$.*

Proof. (a) By Theorems 4.4.4 and 4.4.9,

$$\text{(4.5)} \qquad \lim_{z \to \zeta} |f(z)| = \lim_{z \to \zeta} e^{-g_\Delta(f(z),0)} = \lim_{z \to \zeta} e^{-g_D(z, f^{-1}(0))} = 1,$$

so any continuous extension of f to ζ must satisfy $|f(\zeta)| = 1$.

To show that such an extension exists, we argue by contradiction. Suppose not: then there is a sequence (z_n) in D with $\lim_{n \to \infty} z_n = \zeta$ such that $f(z_{2n}) \to \alpha$ and $f(z_{2n+1}) \to \beta$, for some α, β with $\alpha \neq \beta$. From (4.5) it follows that $|\alpha| = |\beta| = 1$, and multiplying f by a constant, we can further suppose that $\beta = \overline{\alpha}$. Let N be an integer with $2\pi/N < |\alpha - \beta|$, and define

$$u(z) = \log|f^{-1}(z) - \zeta| \quad (z \in \Delta),$$
$$v(z) = \sum_{k=1}^{N} \left(u(e^{2\pi i k/N} z) + u(e^{2\pi i k/N} \overline{z}) \right) \quad (z \in \Delta).$$

Then u, v are subharmonic on Δ, and $v(0) = 2Nu(0) = 2N \log|f^{-1}(0) - \zeta|$. We now seek to estimate this quantity.

Choose a path Γ as in Definition 4.4.12. Given $\epsilon > 0$, there exists t_0 such that $|\Gamma(t) - \zeta| < \epsilon$ for all $t \geq t_0$. Then

$$u \leq \log \epsilon \quad \text{on} \quad S := f(\{\Gamma(t) : t \geq t_0\}),$$

and so

$$v \leq (2N-1) \sup_\Delta u + \log \epsilon \quad \text{on} \quad T := \bigcup_{k=1}^{N} e^{2\pi i k/N}(S \cup S^*),$$

where S^* denotes the reflection of S in the x-axis. Moreover, since $f(\Gamma)$ accumulates at both α and β, the choice of N implies that T separates 0 from $\partial \Delta$. Hence by the maximum principle,

$$v(0) \leq \sup_T v \leq (2N-1) \sup_\Delta u + \log \epsilon.$$

As ϵ is arbitrary, it follows that $v(0) = -\infty$, whence $f^{-1}(0) = \zeta$, contradicting the fact that $f^{-1}(0) \in D$.

4.4. GREEN'S FUNCTIONS

(b) Again, we argue by contradiction. Suppose that $f(\zeta) = f(\zeta') = \alpha$. As both ζ and ζ' are accessible, we can find paths $\Gamma, \Gamma' : [0, \infty) \to D$ such that $\lim_{t \to \infty} \Gamma(t) = \zeta$ and $\lim_{t \to \infty} \Gamma'(t) = \zeta'$. Then $f(\Gamma)$ and $f(\Gamma')$ are two paths in Δ, both ending at α, along which f^{-1} has limits ζ and ζ' respectively. It follows that the function

$$z \mapsto f^{-1}\left(\alpha \frac{z-i}{z+i}\right),$$

which is bounded and holomorphic in the upper half-plane, has distinct asymptotic values ζ, ζ', contradicting Corollary 4.3.12.

(c) By parts (a) and (b), if every boundary point of D is accessible, then f extends to a continuous injection of $\overline{D}$ into $\overline{\Delta}$. A standard compactness argument now shows that $f(\overline{D}) = \overline{\Delta}$, and that f^{-1} is continuous on $\overline{\Delta}$. □

Exercises 4.4

1. Let D be a proper subdomain of $\mathbb{C}_\infty$ such that ∂D is polar. Use Theorem 3.6.9 (a) to show that no Green's function for D exists.

2. Prove that the Green's function of a domain D is (jointly) continuous on $D \times D$. [Hint: use Harnack's inequality.]

3. Let $f: \Delta(0,1) \to \Delta(0,1)$ be a holomorphic function. Prove that

$$\left|\frac{f(z) - f(w)}{1 - f(z)\overline{f(w)}}\right| \leq \left|\frac{z - w}{1 - z\overline{w}}\right| \quad (z, w \in \Delta(0,1)).$$

Show further that, if equality holds for a pair of distinct points z, w, then it holds for all z, w, and deduce that f must be a Möbius map.

4. Let $f: \Delta(0,1) \to \mathbb{C}$ be a holomorphic function such that $f(0) = 0$.

   (i) Prove that if $\operatorname{Re} f(z) \leq 1$ for all z, then

   $$|f(z)| \leq \frac{2|z|}{1 - |z|} \quad (|z| < 1),$$

   and give an example to show that equality can occur. Use this to give another proof of the Borel–Carathéodory inequality (Exercise 1.3.6).

   (ii) Prove that if $|\operatorname{Re} f(z)| \leq 1$ for all z, then

   $$|\operatorname{Im} f(z)| \leq \frac{2}{\pi} \log\left(\frac{1 + |z|}{1 - |z|}\right) \quad (|z| < 1),$$

   and give an example to show that equality can occur.

5. Let D be a domain in $\mathbb{C}_\infty$ such that ∂D is non-polar, and let $w \in D$.

    (i) Show that if u is a negative subharmonic function on D, then
    $$u(z) \leq -C g_D(z,w) \quad (z \in D),$$
    where $C = \Delta u(\{w\})$. [Hint: use the result of Exercise 3.7.3.]

    (ii) Show that if h is a positive harmonic function on $D \setminus \{w\}$ such that $\lim_{z \to \zeta} h(z) = 0$ for n.e. $\zeta \in \partial D$, then
    $$h(z) = C g_D(z,w) \quad (z \in D),$$
    where C is a positive constant. [This time use Exercise 3.7.2.]

6. Show that if a domain D is regular, then so is every domain D' conformally equivalent to D, regardless of how D' is embedded in $\mathbb{C}_\infty$.

7. Let D be a domain in $\mathbb{C}_\infty$ with non-polar boundary, and let E be a compact polar set. Show that
$$g_{D \setminus E}(z,w) = g_D(z,w) \quad (z, w \in D \setminus E).$$

    Use this to give an example to show that Theorem 4.4.10 (b) may fail if the regularity assumption is omitted.

8. Let D be a bounded simply connected domain, and let $f: D \to \Delta$ be a conformal mapping of D onto the unit disc Δ. Suppose that ∂D is *analytic*, i.e. for each $\zeta \in \partial D$ there is a neighbourhood N of ζ and a conformal mapping ϕ of N onto Δ such that $\phi(N \cap D) = \{z \in \Delta : \operatorname{Im} z > 0\}$. Use the technique of the reflection principle (Theorem 1.2.9) to prove that f extends to a conformal map of a neighbourhood of $\overline{D}$ to a neighbourhood of $\overline{\Delta}$.

4.5 The Poisson–Jensen Formula

If u is a subharmonic function on a domain containing a closed disc $\overline{\Delta}$, then we saw in Theorem 2.4.1 that $u \leq P_\Delta u$ on Δ. The difference $P_\Delta u - u$ measures how far u is from being harmonic on Δ, and one would expect this to depend on the size of the generalized Laplacian Δu. The following theorem not only makes this precise, but extends it to a wide range of other domains. It is the culmination of a whole sequence of earlier results.

4.5. THE POISSON–JENSEN FORMULA

Theorem 4.5.1 (Poisson–Jensen Formula) *Let D be a bounded regular domain in $\mathbb{C}$, and let u be a function subharmonic on a neighbourhood of $\overline{D}$, with $u \not\equiv -\infty$ on D. Then*

$$u(z) = \int_{\partial D} u(\zeta)\, d\omega_D(z,\zeta) - \frac{1}{2\pi} \int_D g_D(z,w)\, \Delta u(w) \quad (z \in D).$$

Proof. We begin with the claim that, if $z \in D$, then

$$(4.6) \quad \int_{\partial D} \log|\zeta - w|\, d\omega_D(z,\zeta) = \begin{cases} \log|z-w| + g_D(z,w), & w \in D, \\ \log|z-w|, & w \in \mathbb{C} \setminus D. \end{cases}$$

If $w \in D$, this follows directly from Theorem 4.4.7. If $w \in \mathbb{C} \setminus \overline{D}$, then the function $z' \mapsto \log|z' - w|$ is harmonic on a neighbourhood of $\overline{D}$, and so in this case (4.6) follows from the definition of harmonic measure. Finally, suppose that $w \in \partial D$. Then as D is connected, Theorem 3.8.3 gives

$$\begin{aligned}
\int_{\partial D} \log|\zeta - w|\, d\omega_D(z,\zeta) &= \limsup_{\substack{w' \to w \\ w' \in D}} \left(\int_{\partial D} \log|\zeta - w'|\, d\omega_D(z,\zeta) \right) \\
&= \limsup_{\substack{w' \to w \\ w' \in D}} (\log|z-w'| + g_D(z,w')) \\
&= \log|z-w| + \lim_{\substack{w' \to w \\ w' \in D}} g_D(w',z) \\
&= \log|z-w|,
\end{aligned}$$

the last equality coming from Theorem 4.4.9 and the hypothesis that D is a regular domain. Thus (4.6) holds in this case too, and the claim is justified.

Now choose a bounded domain D_1 containing $\overline{D}$ such that u is subharmonic on a neighbourhood of $\overline{D_1}$. By the Riesz decomposition theorem (Theorem 3.7.9) we can write $u = p_\mu + h$ on D_1, where $\mu = (2\pi)^{-1} \Delta u|_{D_1}$ and h is harmonic on D_1. Then for $z \in D$, we have

$$\begin{aligned}
\int_{\partial D} &u(\zeta)\, d\omega_D(z,\zeta) \\
&= \int_{\partial D} \left(\int_{D_1} \log|\zeta - w|\, d\mu(w) \right) d\omega_D(z,\zeta) + \int_{\partial D} h(\zeta)\, d\omega_D(z,\zeta) \\
&= \int_{D_1} \left(\int_{\partial D} \log|\zeta - w|\, d\omega_D(z,\zeta) \right) d\mu(w) + h(z) \\
&= \int_D g_D(z,w)\, d\mu(w) + \int_{D_1} \log|z-w|\, d\mu(w) + h(z) \\
&= \frac{1}{2\pi} \int_D g_D(z,w)\, \Delta u(w) + u(z).
\end{aligned}$$

Rearranging this equation gives the result. □

As a special case, we recapture the classical Poisson–Jensen formula for holomorphic functions on a disc, used, for example, in value-distribution theory.

Corollary 4.5.2 *Let f be a function holomorphic on a neighbourhood of $\overline{\Delta}(0,1)$, with $f \not\equiv 0$. Then*

$$\log|f(z)| = \frac{1}{2\pi}\int_0^{2\pi} \frac{1-|z|^2}{|e^{i\theta}-z|^2} \log|f(e^{i\theta})|\,d\theta - \sum_{j=1}^n \log\left|\frac{1-z\overline{w}_j}{z-w_j}\right| \quad (|z|<1),$$

where $w_1, \ldots, w_n$ are the zeros of f in $\Delta(0,1)$, counted according to multiplicity.

Proof. Set $\Delta = \Delta(0,1)$, and recall that

$$d\omega_\Delta(z, e^{i\theta}) = \frac{1}{2\pi}\frac{1-|z|^2}{|e^{i\theta}-z|^2}\,d\theta \quad \text{and} \quad g_\Delta(z,w) = \log\left|\frac{1-z\overline{w}}{z-w}\right|.$$

Also, by Theorem 3.7.8, $\Delta(\log|f|)$ consists of 2π-masses at the zeros of f. The result follows by feeding these facts into Theorem 4.5.1. $\square$

If we merely suppose that u is subharmonic on D, rather than on a neighbourhood of $\overline{D}$, then Δu may be an infinite measure on D, and it is no longer clear whether the integral $\int_D g_D(z,w)\Delta u(w)$ converges. This turns out to depend on whether u has a harmonic majorant, a concept which we now define.

Definition 4.5.3 *Let u be a subharmonic function on a domain D. A harmonic majorant of u is a harmonic function h on D such that $h \geq u$ there. If also $h \leq k$, for every other harmonic majorant k of u, then h is called the least harmonic majorant of u.*

Theorem 4.5.4 *Let D be a subdomain of $\mathbb{C}$ such that ∂D is non-polar, and let u be a subharmonic function on D with $u \not\equiv -\infty$.*

(a) *If u has a harmonic majorant on D, then it has a least one, h, and*

$$u(z) = h(z) - \frac{1}{2\pi}\int_D g_D(z,w)\,\Delta u(w) \quad (z \in D).$$

(b) *If u has no harmonic majorant on D, then*

$$\frac{1}{2\pi}\int_D g_D(z,w)\,\Delta u(w) = \infty \quad (z \in D).$$

4.5. THE POISSON–JENSEN FORMULA

Proof. Let $(D_n)_{n \geq 1}$ be a sequence of relatively compact subdomains of D such that $D_1 \subset D_2 \subset D_3 \subset \cdots$ and $\cup_n D_n = D$. We can further arrange that each component of $\mathbb{C}_\infty \setminus D_n$ contains at least two points, so that by Theorem 4.2.2 D_n is a regular domain. For $n \geq 1$, define

$$h_n(z) = \int_{\partial D_n} u(\zeta)\, d\omega_{D_n}(z, \zeta) \quad (z \in D_n),$$

so that h_n is a harmonic function on D_n. Then by Theorem 4.5.1

$$u(z) = h_n(z) - \frac{1}{2\pi} \int_{D_n} g_{D_n}(z, w)\, \Delta u(w) \quad (z \in D_n).$$

Since $g_{D_n} \uparrow g_D$ as $n \to \infty$, it follows that $h_n \uparrow h$ on D, where h satisfies

(4.7) $\qquad u(z) = h(z) - \dfrac{1}{2\pi} \displaystyle\int_D g_D(z, w)\, \Delta u(w) \quad (z \in D).$

Also, by Harnack's theorem (Theorem 1.3.9), either h is harmonic on D or $h \equiv \infty$ there. We now consider two cases.

Suppose first that u has a harmonic majorant k on D. Then for each n, it follows from the definition of h_n that

$$h_n(z) \leq \int_{\partial D_n} k(\zeta)\, d\omega_{D_n}(z, \zeta) = k(z) \quad (z \in D_n),$$

and hence $h \leq k$ on D. In particular $h \not\equiv \infty$, so h must be harmonic on D. Equation (4.7) then shows that h is a harmonic majorant of u, and we have just seen that it is less than any other harmonic majorant k, so it must be the least one. This completes the proof of (a).

Now suppose that u has no harmonic majorant on D. Then h cannot be harmonic, for otherwise it would be such a majorant. Consequently $h \equiv \infty$, and we conclude from (4.7) that

$$\frac{1}{2\pi} \int_D g_D(z, w)\, \Delta u(w) = \infty \quad (z \in D).$$

This completes the proof of (b). □

This has an interesting consequence for holomorphic functions.

Corollary 4.5.5 *Let D be a domain in $\mathbb{C}$ such that ∂D is non-polar, let f be a holomorphic function on D, and let z_0 be a point in D such that $f(z_0) \neq 0$. Then $\log |f|$ has a harmonic majorant on D if and only if*

$$\sum_j g_D(z_0, w_j) < \infty,$$

where $w_1, w_2, w_3, \ldots$ are the zeros of f. In particular, this series must converge if f is bounded.

Proof. If we write $u = \log|f|$, then Δu consists of 2π-masses at the zeros of f, and so
$$\int_D g_D(z,w)\,\Delta u(w) = \sum_j g_D(z,w_j).$$
The result therefore follows directly from Theorem 4.5.4. $\square$

The *Carleman–Milloux problem*, which arises naturally out of the problem of estimating harmonic measure, is to find the best upper bound for a subharmonic function u on $\Delta(0,1)$ satisfying
$$(4.8) \qquad \sup_{|z|=r} u(z) \le 0 \quad \text{and} \quad \inf_{|z|=r} u(z) \le -1 \qquad (0 \le r < 1).$$
This can be regarded as a quantitative version of Exercise 3.8.3. As an application of the 'Green machine', we now present the beautiful solution found by Beurling and Nevanlinna.

Theorem 4.5.6 (Beurling–Nevanlinna Theorem) *Let u be a subharmonic function on $\Delta(0,1)$ satisfying (4.8). Then*
$$u(z) \le -\frac{2}{\pi}\sin^{-1}\left(\frac{1-|z|}{1+|z|}\right) \qquad (|z| < 1),$$
and this bound is sharp.

The proof relies on two lemmas. The first of these is an elementary inequality for the Green's function on the disc.

Lemma 4.5.7 *If $\Delta = \Delta(0,1)$, then*
$$g_\Delta(-|z|, |w|) \le g_\Delta(z,w) \le g_\Delta(|z|, |w|) \qquad (z, w \in \Delta).$$

Proof. Let $z, w \in \Delta$, and write $z = |z|e^{i\alpha}$ and $w = |w|e^{i\beta}$. Then
$$\left|\frac{1-z\overline{w}}{z-w}\right|^2 = 1 + \frac{(1-|z|^2)(1-|w|^2)}{|z|^2 + |w|^2 - 2|z||w|\cos(\alpha-\beta)},$$
which is maximized when $\cos(\alpha-\beta) = 1$, and minimized when $\cos(\alpha-\beta) = -1$. Hence
$$\left|\frac{1+|z||w|}{|z|+|w|}\right|^2 \le \left|\frac{1-z\overline{w}}{z-w}\right|^2 \le \left|\frac{1-|z||w|}{|z|-|w|}\right|^2,$$
and since we know that
$$g_\Delta(z,w) = \log\left|\frac{1-z\overline{w}}{z-w}\right|,$$
the corresponding inequality also holds for g_Δ. $\square$

4.5. THE POISSON–JENSEN FORMULA

Lemma 4.5.8 *Let $\Delta = \Delta(0,1)$ and $I = [0,1)$, and define v on Δ by*

$$v(z) = \begin{cases} -\omega_{\Delta \setminus I}(z, I), & z \in \Delta \setminus I, \\ -1, & z \in I. \end{cases}$$

Then v is subharmonic on Δ and harmonic on $\Delta \setminus I$. Also

$$v(z) = -\frac{1}{2\pi} \int_I g_\Delta(z,w) \, \Delta v(w) \quad (z \in \Delta),$$

and

$$v(-x) = -\frac{2}{\pi} \sin^{-1}\left(\frac{1-x}{1+x}\right) \quad (x \in I).$$

Proof. By Theorem 4.3.4 v is harmonic on $\Delta \setminus I$, and by the gluing theorem it follows that v is subharmonic on Δ.

Clearly 0 is a harmonic majorant of v. In fact it is the least one, for if k is another, then

$$\limsup_{z \to \zeta} -k(z) \leq \lim_{z \to \zeta} -v(z) = 0 \quad (\zeta \in \partial\Delta \setminus \{1\}),$$

and so by the extended maximum principle $k \geq 0$ on Δ. Applying Theorem 4.5.4, we deduce that

$$v(z) = 0 - \frac{1}{2\pi} \int_I g_\Delta(z,w) \, \Delta v(w) \quad (z \in \Delta),$$

the integral being taken over I since $\Delta v = 0$ on $\Delta \setminus I$.

Lastly, computing the harmonic measure via conformal mapping, we have

$$\omega_{\Delta \setminus I}(z, I) = 1 - \frac{2}{\pi} \arg\left(\frac{1+\sqrt{z}}{1-\sqrt{z}}\right) \quad (z \in \Delta \setminus I),$$

and the final equation in the lemma follows easily from this. □

Proof of Theorem 4.5.6. Let $\Delta = \Delta(0,1)$ and $U = \{z \in \Delta : u(z) < -1\}$. We can suppose that $\inf_{|z|=r} u(z) < -1$ for all r: otherwise work with $u - \epsilon$, and ultimately let $\epsilon \to 0$. Thus if we define $T: \Delta \to I$ by $T(z) = |z|$, then $T(U) = I$.

Let v be the function defined in Lemma 4.5.8. Given $\rho < 1$, we can find a compact subset K of U such that $T(K) = [0, \rho]$. Then by Theorem A.4.4 of the Appendix, there exists a finite Borel measure μ on K such that $\mu T^{-1} = \Delta v|_{[0,\rho]}$. Define a function h on Δ by

$$h(z) = -\frac{1}{2\pi} \int_K g_\Delta(z,w) \, d\mu(w) \quad (z \in \Delta).$$

Then h is harmonic on $\Delta \setminus K$, and $\lim_{z \to \zeta} h(z) = 0$ for all $\zeta \in \partial \Delta$. Also if $z \in \Delta$, then using the right-hand inequality of Lemma 4.5.7, we get

$$\begin{aligned} h(z) &\geq -\frac{1}{2\pi} \int_K g_\Delta(|z|, |w|) \, d\mu(w) \\ &= -\frac{1}{2\pi} \int_{[0,\rho]} g_\Delta(|z|, w) \, \Delta v(w) \\ &\geq -\frac{1}{2\pi} \int_I g_\Delta(|z|, w) \, \Delta v(w) \\ &= v(|z|) = -1. \end{aligned}$$

Hence if $\zeta \in \partial(\Delta \setminus K)$, then

$$\limsup_{\substack{z \to \zeta \\ z \in \Delta \setminus K}} (u - h)(z) \leq \left\{ \begin{array}{ll} 0, & \zeta \in \partial \Delta \\ u(\zeta) - (-1), & \zeta \in \partial K \end{array} \right\} \leq 0,$$

and so by the maximum principle $u \leq h$ on $\Delta \setminus K$. Since also $u \leq -1 \leq h$ on K, we in fact have $u \leq h$ on the whole of Δ. Applying now the left-hand inequality of Lemma 4.5.7, we deduce that, for each $z \in \Delta$,

$$\begin{aligned} u(z) &\leq -\frac{1}{2\pi} \int_K g_\Delta(-|z|, |w|) \, d\mu(w) \\ &= -\frac{1}{2\pi} \int_{[0,\rho]} g_\Delta(-|z|, w) \, \Delta v(w). \end{aligned}$$

As this holds for each $\rho < 1$, we can let $\rho \to 1$ to obtain

$$\begin{aligned} u(z) &\leq -\frac{1}{2\pi} \int_I g_\Delta(-|z|, w) \, \Delta v(w) \\ &= v(-|z|) = -\frac{2}{\pi} \sin^{-1}\left(\frac{1-|z|}{1+|z|}\right), \end{aligned}$$

the final equality coming from Lemma 4.5.8. This proves the desired bound for u.

To show that this bound is sharp, we note that for each θ, the function $u_\theta(z) := v(e^{i\theta} z)$ satisfies the hypotheses of the theorem, and so any general upper bound for $u(z)$ must be at least as large as

$$\sup_\theta u_\theta(z) = \sup_\theta v(e^{i\theta} z) = v(-|z|) = -\frac{2}{\pi} \sin^{-1}\left(\frac{1-|z|}{1+|z|}\right).$$

With this observation, the proof is complete. □

4.5. THE POISSON–JENSEN FORMULA

As a consequence of this result, we can derive some general estimates for harmonic measure on a simply connected domain.

Corollary 4.5.9 *Let D be a simply connected subdomain of $\mathbb{C}$ such that $0 \notin D$, and let $\rho > 0$.*

(a) *If $z \in D$ and $|z| < \rho$, then*
$$\omega_D\bigl(z, \partial D \cap \Delta(0,\rho)\bigr) \geq \frac{2}{\pi} \sin^{-1}\left(\frac{\rho - |z|}{\rho + |z|}\right).$$

(b) *If $z \in D$ and $|z| > \rho$, then*
$$\omega_D\bigl(z, \partial D \cap \Delta(0,\rho)\bigr) \leq \frac{2}{\pi} \cos^{-1}\left(\frac{|z| - \rho}{|z| + \rho}\right).$$

Proof. (a) Define u on $\Delta(0, \rho)$ by
$$u(z) = \begin{cases} -\omega_D\bigl(z, \partial D \cap \Delta(0,\rho)\bigr), & z \in \Delta(0,\rho) \cap D, \\ -1, & z \in \Delta(0,\rho) \setminus D. \end{cases}$$

As D is simply connected, Theorem 4.2.1 guarantees that it is a regular domain, and hence the gluing theorem applies to show u is subharmonic on $\Delta(0, \rho)$. Evidently $u \leq 0$. Also no circle $|z| = r$ can be entirely contained in D, for then it would separate 0 and ∞, both of which lie outside D, contradicting the fact that D is simply connected. Hence
$$\inf_{|z|=r} u(z) = -1 \quad (0 \leq r < \rho).$$

Applying Theorem 4.5.6 to the function $z' \mapsto u(\rho z')$ on $\Delta(0,1)$, we deduce that
$$u(z) \leq -\frac{2}{\pi} \sin^{-1}\left(\frac{1 - |z/\rho|}{1 + |z/\rho|}\right) \quad (z \in \Delta(0,\rho)),$$
which gives the result.

(b) Let D' be the image of D under the inversion $z \mapsto 1/z$. Then if $z \in D$,
$$\begin{aligned}\omega_D\bigl(z, \partial D \cap \Delta(0,\rho)\bigr) &= \omega_{D'}\bigl(1/z, \partial D' \setminus \overline{\Delta}(0, 1/\rho)\bigr) \\ &\leq 1 - \omega_{D'}\bigl(1/z, \partial D' \cap \Delta(0, 1/\rho)\bigr).\end{aligned}$$

Applying part (a) to D', it follows that if also $|z| > \rho$ then
$$\begin{aligned}\omega_D\bigl(z, \partial D \cap \Delta(0,\rho)\bigr) &\leq 1 - \frac{2}{\pi} \sin^{-1}\left(\frac{1/\rho - 1/|z|}{1/\rho + 1/|z|}\right) \\ &= \frac{2}{\pi} \cos^{-1}\left(\frac{|z| - \rho}{|z| + \rho}\right).\end{aligned}$$

This completes the proof. □

Exercises 4.5

1. Let D be a bounded regular domain in $\mathbb{C}$, let μ be a finite Borel measure such that $\operatorname{supp}\mu \subset D$, and let $\phi\colon \partial D \to \mathbb{R}$ be a continuous function. Show that if
$$u(z) = \int_{\partial D} \phi(\zeta)\, d\omega_D(z,\zeta) - \frac{1}{2\pi}\int_D g_D(z,w)\, d\mu(w) \quad (z \in D),$$
then
(4.9)
$$\begin{cases} \Delta u = \mu & \text{on } D, \\ u \to \phi & \text{on } \partial D, \end{cases}$$
and prove that u is the unique subharmonic function on D satisfying (4.9). [Hint: apply the Poisson–Jensen formula to p_μ.]

2. Let u be a subharmonic function on $\Delta(0,1)$. Show that u has a harmonic majorant if and only if
$$\sup_{r<1}\left(\frac{1}{2\pi}\int_0^{2\pi} u(re^{it})\, dt\right) < \infty.$$

3. Let D be a domain in $\mathbb{C}_\infty$ such that ∂D is polar. Show that a subharmonic function on D has a harmonic majorant if and only if it is itself harmonic.

4. Let f be a holomorphic function on a domain D, and let $w_1, w_2, w_3, \ldots$ be the zeros of f.

   (i) Show that if $D = \{z : |z| < 1\}$, then $\log|f|$ has a harmonic majorant if and only if
   $$\sum_j (1 - |w_j|) < \infty.$$

   (ii) Show that if $D = \{z : \operatorname{Re} z > 0\}$, then $\log|f|$ has a harmonic majorant if and only if
   $$\sum_j \frac{\operatorname{Re} w_j}{1 + |w_j|^2} < \infty.$$

5. Let $\Delta = \Delta(0,1)$, and let F be a closed subset of Δ such that $\Delta \setminus F$ is connected. Use the technique of the proof of Theorem 4.5.6 to show that
$$\omega_{\Delta\setminus F}(z, \partial F) \geq \omega_{\Delta\setminus F^*}(-|z|, F^*) \quad (z \in \Delta \setminus F),$$
where $F^* = \{|z| : z \in F\}$.

4.5. THE POISSON–JENSEN FORMULA

6. Let D be a simply connected domain in $\mathbb{C}$ such that $0 \notin D$, and let $z \in D$. Prove that
$$\omega_D(z, \partial D \cap \Delta(0,\rho)) = O(\rho^{1/2}) \quad \text{as } \rho \to 0,$$
and give an example to show the exponent $1/2$ is sharp.

7. Let $(u_n)_{n \geq 1}$ be a sequence of negative subharmonic functions on $\Delta(0,1)$ such that $u_n(0) \geq -C > -\infty$ for all n.

    (i) Fix ρ with $0 < \rho < 1$. Show that each measure $\mu_n := \Delta u_n$ satisfies
    $$\mu_n(\overline{\Delta}(0,\rho)) \leq 2\pi C / \log(1/\rho),$$
    and deduce that there is a subsequence (μ_{n_j}) which is weak*-convergent on $\overline{\Delta}(0,\rho)$ to some finite measure μ.

    (ii) Define v_n on $\Delta(0,\rho)$ by
    $$v_n(z) = \frac{1}{2\pi} \int_{\Delta(0,\rho)} g_{\Delta(0,\rho)}(z,w) \, d\mu_n(w),$$
    and let v be the corresponding function for μ. Show that
    $$\liminf_{j \to \infty} v_{n_j} \geq v \quad \text{on } \Delta(0,\rho).$$

    (iii) Show that if $0 < r < \rho$, then for each n,
    $$\int_0^{2\pi} v_n(re^{it}) \, dt = - \int_{\Delta(0,\rho)} \max\bigl(\log|w/\rho|, \log(r/\rho)\bigr) \, d\mu_n(w),$$
    with a corresponding equation for v, and hence deduce that
    $$\int_0^{2\pi} v_{n_j}(re^{it}) \, dt \to \int_0^{2\pi} v(re^{it}) \, dt.$$

    (iv) Show that $u_n = h_n - v_n$ on $\Delta(0,\rho)$, where each h_n is a negative harmonic function on $\Delta(0,\rho)$. Deduce that there exists a further subsequence (which, by relabelling, we may also call (h_{n_j})), which is locally uniformly convergent on $\Delta(0,\rho)$ to some harmonic function h.

    (v) Set $u = h - v$. Show that u is subharmonic on $\Delta(0,\rho)$, and combine the results of the three previous parts to prove that
    $$\int_0^{2\pi} |u_{n_j}(re^{it}) - u(re^{it})| \, dt \to 0 \quad (0 < r < \rho).$$

(vi) Conclude, using a diagonal argument, that there exists a subsequence (u_{n_j}) and a subharmonic function u on $\Delta(0,1)$ such that
$$\int_0^{2\pi} |u_{n_j}(re^{it}) - u(re^{it})|\,dt \to 0 \quad (0 < r < 1).$$
Show also that $\int_0^{2\pi} |u_n(re^{it})|\,dt \le 2\pi C$ for all r, n, and hence deduce that
$$\int_{\Delta(0,1)} |u_{n_j}(z) - u(z)|\,dA(z) \to 0.$$

Notes on Chapter 4

§4.1

The definition of barrier (Definition 4.1.4) is a little weaker than is customary. Most authors insist, in addition, that b be bounded away from 0 outside each neighbourhood of ζ_0. This allows one to shorten the proof of Theorem 4.1.5, since Lemma 4.1.7 can then be omitted. However, even though it may cost some extra effort initially, our definition saves work in the long run, because it is then much easier to construct barriers in specific instances (see e.g. Theorems 4.2.1 and 4.2.4).

§4.3

For more information about estimation of harmonic measure and its applications, see the book of Garnett [32] and the references cited therein.

§4.4

Theorem 4.4.13 and Exercise 8 illustrate the close relationship between the smoothness of the boundary of a simply connected domain and the smoothness up to the boundary of the Riemann mapping on that domain. For many further results of this type we refer to Pommerenke's book [51].

Exercises 3 and 5 are two generalizations of Schwarz's lemma. Dineen's book [25] discusses the far-reaching implications of these ideas.

§4.5

The proof of Theorem 4.5.6 is based on [44, Chapter IV, §5], as is the slight generalization given in Exercise 5.

The subsequence principle for subharmonic functions, outlined in Exercise 7, is from [3]. See [26] for a recent application to complex dynamics.

Chapter 5

Capacity

5.1 Capacity as a Set Function

Even though polar sets have played a prominent rôle in the theory developed so far, we still lack an effective means of determining whether or not a given set is polar. Thus it was only by a very indirect method that we were able to demonstrate the existence of uncountable polar sets in Section 3.5, and nothing we have yet proved will tell us whether, for example, the Cantor set is polar.

More generally, it is desirable to be able to gauge, in some way, how close a set is to being polar. In the case of a compact set, the energy $I(\nu)$ of its equilibrium measure, a quantity that has already cropped up several times, provides just such an indicator. Taking exponentials in order to make it positive, we are led to the following definition.

Definition 5.1.1 The *logarithmic capacity* of a subset E of $\mathbb{C}$ is given by

$$c(E) := \sup_{\mu} e^{I(\mu)},$$

where the supremum is taken over all Borel probability measures μ on $\mathbb{C}$ whose support is a compact subset of E. In particular, if K is a compact set with equilibrium measure ν, then

$$c(K) = e^{I(\nu)}.$$

Here it is understood that $e^{-\infty} = 0$, so that $c(E) = 0$ precisely when E is polar. There are several other capacities with this property, but the logarithmic capacity enjoys the advantage of particularly close links with complex analysis. Since it is the only one we shall study, it will henceforth be referred to simply as 'the capacity'.

We begin by listing some of its elementary properties.

Theorem 5.1.2 (a) *If $E_1 \subset E_2$ then $c(E_1) \leq c(E_2)$.*

(b) *If $E \subset \mathbb{C}$ then $c(E) = \sup\{c(K) :$ compact $K \subset E\}$.*

(c) *If $E \subset \mathbb{C}$ then $c(\alpha E + \beta) = |\alpha| c(E)$ for all $\alpha, \beta \in \mathbb{C}$.*

(d) *If K is a compact subset of $\mathbb{C}$ then $c(K) = c(\partial_e K)$.*

Proof. Both (a) and (b) follow immediately from Definition 5.1.1.

To prove (c), let $T: \mathbb{C} \to \mathbb{C}$ be the map $T(z) = \alpha z + \beta$. Then $\operatorname{supp} \mu \subset E$ if and only if $\operatorname{supp} \mu T^{-1} \subset \alpha E + \beta$, and

$$I(\mu T^{-1}) = I(\mu) + \log |\alpha|.$$

The result follows easily from this.

Finally, (d) is a direct consequence of Theorem 3.7.6. □

Since capacity is a monotone set function, it is natural to ask if it is continuous with respect to increasing or decreasing sequences. The following result gives an answer.

Theorem 5.1.3 (a) *If $K_1 \supset K_2 \supset K_3 \supset \cdots$ are compact subsets of $\mathbb{C}$ and $K = \cap_n K_n$, then*

$$c(K) = \lim_{n \to \infty} c(K_n).$$

(b) *If $B_1 \subset B_2 \subset B_3 \subset \cdots$ are Borel subsets of $\mathbb{C}$ and $B = \cup_n B_n$, then*

$$c(B) = \lim_{n \to \infty} c(B_n).$$

Proof. (a) By Theorem 5.1.2 (a) we certainly have

(5.1) $$c(K_1) \geq c(K_2) \geq \cdots \geq c(K).$$

In the other direction, for each $n \geq 1$ let ν_n be an equilibrium measure for K_n. Then $\nu_n \in \mathcal{P}(K_1)$ for all n, so by Theorem A.4.2 there is a subsequence (ν_{n_k}) which is weak*-convergent to some $\nu \in \mathcal{P}(K_1)$. Applying Lemma 3.3.3, we deduce that

$$\limsup_{k \to \infty} I(\nu_{n_k}) \leq I(\nu).$$

Moreover, since $\operatorname{supp} \nu_n \subset K_n$ for all n, it follows that $\operatorname{supp} \nu \subset K$, and so $e^{I(\nu)} \leq c(K)$. Thus we obtain

$$\limsup_{k \to \infty} c(K_{n_k}) \leq c(K),$$

and combining this with (5.1) yields the desired conclusion.

5.1. CAPACITY AS A SET FUNCTION

(b) Again using Theorem 5.1.2 (a), we have

(5.2) $$c(B_1) \leq c(B_2) \leq \cdots \leq c(B).$$

In the other direction, let K be a compact subset of B, and let ν be an equilibrium measure for K. Since $\nu(B_n \cap K) \to \nu(K) = 1$ as $n \to \infty$, we can use Theorem A.2.2 to produce compact sets $K_n \subset B_n \cap K$ such that $K_1 \subset K_2 \subset \cdots$ and $\nu(K_n) \to 1$. For n sufficiently large, we have $\nu(K_n) > 0$, and for these n we define

$$\mu_n = \frac{\nu|K_n}{\nu(K_n)}.$$

Then μ_n is a Borel probability measure on K_n, and

$$I(\mu_n) = \frac{1}{\nu(K_n)^2} \int_K \int_K \log|z - w| 1_{K_n}(z) 1_{K_n}(w) \, d\nu(z) \, d\nu(w).$$

As $n \to \infty$, we have $\nu(K_n) \to 1$ and $1_{K_n} \uparrow 1_K$ ν-almost everywhere, so

$$\lim_{n \to \infty} I(\mu_n) = \int_K \int_K \log|z - w| \, d\nu(z) \, d\nu(w) = I(\nu).$$

Since each μ_n is supported on a compact subset of B_n, we have $c(B_n) \geq e^{I(\mu_n)}$, and it follows that

$$\liminf_{n \to \infty} c(B_n) \geq c(K).$$

Finally, as K is an arbitrary subset of B, Theorem 5.1.2 (b) implies that

$$\liminf_{n \to \infty} c(B_n) \geq c(B),$$

and combining this with (5.2), we again obtain the desired conclusion. □

Theorem 5.1.3 (a) is false for general Borel sets, indeed even for bounded open sets. For example, consider the sequence

$$U_n := \{z \in \mathbf{C} : -1 < \operatorname{Re} z < 1, \ 0 < \operatorname{Im} z < 1/n\} \quad (n \geq 1).$$

Then clearly $U_1 \supset U_2 \supset \cdots$ and $\bigcap_n U_n = \emptyset$. But also each set U_n contains a translate of the non-polar set $[0, 1]$, and so $c(U_n) \geq c([0, 1]) > 0$ for all n.

However, it can be shown that, given a bounded Borel set B, we have

(5.3) $$c(B) = \inf\{c(U) : \text{ open } U \supset B\}.$$

This result, due to Choquet, looks like the dual to Theorem 5.1.2 (b), but actually it lies much deeper, and we shall not prove it here.

Capacity is not an additive set-function, like a measure. For example, the unit disc $\overline{\Delta}(0,1)$, which has finite capacity, contains infinitely many disjoint translates of the unit interval $[0,1]$, which has strictly positive capacity since it is non-polar. There is however a relation between capacity and unions.

Theorem 5.1.4 *Let (B_n) be a (finite or infinite) sequence of Borel subsets of $\mathbb{C}$, let $B = \cup_n B_n$, and let $d > 0$.*

(a) *If* $\operatorname{diam}(B) \leq d$, *then* $c(B) \leq d$ *and*

$$(5.4) \qquad \frac{1}{\log(d/c(B))} \leq \sum_n \frac{1}{\log(d/c(B_n))}.$$

(b) *If* $\operatorname{dist}(B_j, B_k) \geq d$ *whenever* $j \neq k$, *then*

$$(5.5) \qquad \frac{1}{\log^+(d/c(B))} \geq \sum_n \frac{1}{\log^+(d/c(B_n))}.$$

Here, we interpret $1/0$ as ∞, and $1/\infty$ as 0. Thus, for example, part (a) re-proves the result that a countable union of Borel polar sets is polar (at least provided the union is bounded, but the unbounded case can then be deduced using Theorem 5.1.3 (b)).

Proof. (a) We begin by noting that if $\operatorname{diam}(B) \leq d$ then, for any probability measure μ compactly supported on B, we have

$$I(\mu) = \int_B \int_B \log|z-w|\, d\mu(z)\, d\mu(w) \leq \int_B \int_B (\log d)\, d\mu(z)\, d\mu(w) = \log d,$$

and therefore $c(B) \leq d$, as claimed.

It is enough to prove (5.4) in the case where there are just two sets B_1, B_2. The case for n sets then follows by induction, and for infinitely many sets the result can be deduced using Theorem 5.1.3 (b). By scaling, we can also suppose that $d = 1$.

Let K be a compact subset of B, and let $\epsilon > 0$. Our aim is to show that

$$(5.6) \qquad \frac{1-\epsilon}{\log(1/c(K))} \leq \frac{1}{\log(1/c(B_1))} + \frac{1}{\log(1/c(B_2))}.$$

This inequality is clear if $c(K) = 0$, so we may as well assume $c(K) > 0$. In that case $I(\nu) > -\infty$, where ν is the equilibrium measure for K. Since $\nu(B_1 \cap K) + \nu(B_2 \cap K) \geq \nu(K) = 1$, we can apply Theorem A.2.2 to obtain compact sets $K_j \subset B_j \cap K$ $(j = 1, 2)$ such that

$$\nu(K_1) + \nu(K_2) > 1 - \epsilon.$$

5.1. CAPACITY AS A SET FUNCTION

For $j = 1, 2$, let ν_j be an equilibrium measure for K_j. Then we have

$$I(\nu) \leq \int_{K_j} p_\nu \, d\nu_j = \int_K p_{\nu_j} \, d\nu \leq \int_{K_j} p_{\nu_j} \, d\nu = I(\nu_j)\nu(K_j).$$

Here, the first relation holds because $p_\nu \geq I(\nu)$ on $\mathbb{C}$, the second comes from Fubini's theorem, the third is true because $p_{\nu_j} \leq 0$ on K (recall that $\operatorname{diam}(K) \leq d = 1$), and the fourth holds since $p_{\nu_j} = I(\nu_j)$ nearly everywhere (and hence ν-almost everywhere) on K_j. Now $I(\nu) = \log c(K) \leq 0$ since $\operatorname{diam}(K) \leq d = 1$, and likewise for $I(\nu_j)$, so we obtain

$$\frac{\nu(K_j)}{\log(1/c(K))} \leq \frac{1}{\log(1/c(K_j))} \leq \frac{1}{\log(1/c(B_j))} \quad (j = 1, 2).$$

Summing over j gives (5.6). Finally, letting $\epsilon \to 0$ in (5.6), and taking the supremum over all compact $K \subset B$, we obtain (5.4).

(b) As in (a), we can suppose that there are just two sets B_1, B_2, and that $d = 1$.

Let K_1, K_2 be compact subsets of B_1, B_2 respectively. This time, our aim is to show that

(5.7) $$\frac{1}{\log^+(1/c(B))} \geq \frac{1}{\log^+(1/c(K_1))} + \frac{1}{\log^+(1/c(K_2))}.$$

We can assume that $0 < c(K_j) \leq c(B) < 1$, since otherwise (5.7) is clear anyway. For $j = 1, 2$, let ν_j be the equilibrium measure for K_j, and set $\mu = (1-t)\nu_1 + t\nu_2$, where

$$t = \frac{I(\nu_1)}{I(\nu_1) + I(\nu_2)}.$$

Since $-\infty < I(\nu_j) < 0$, it follows that $0 < t < 1$, and hence μ is a probability measure with

$$I(\mu) \geq (1-t)^2 I(\nu_1) + t^2 I(\nu_2) = \frac{I(\nu_1) I(\nu_2)}{I(\nu_1) + I(\nu_2)}.$$

Now μ is supported on $K_1 \cup K_2 \subset B$, so $I(\mu) \leq \log c(B)$, and hence

$$\log c(B) \geq \frac{\log c(K_1) \log c(K_2)}{\log c(K_1) + \log c(K_2)}.$$

Since $\log c(B)$ and $\log c(K_j)$ are all negative, when this inequality is inverted it becomes (5.7). Finally, taking the supremum in (5.7) over all compact $K_1 \subset B_1$ and $K_2 \subset B_2$, we obtain (5.5). □

We conclude by mentioning that capacity can behave badly with respect to complements. An example appears in Exercise 3 below.

Exercises 5.1

1. Evaluate $I(\nu)$, where ν is normalized Lebesgue measure on the unit circle, and hence show that $c(\overline{\Delta}(0,1)) = 1$.

2. Show that there is no constant γ such that
$$c(E_1 \cup E_2) \leq \gamma(c(E_1) + c(E_2)) \quad (E_1, E_2 \subset \mathbb{C}).$$

3. (i) Show that, given $E \subset \mathbb{C}$, there exists an F_σ subset F of E such that
$$c(F) = c(E).$$

   (ii) Let S be a subset of $[0,1]$ which is not an F_σ. Show that every F_σ subset F of $[0,1] \times S$ satisfies
$$c(([0,1] \times S) \setminus F) \geq c([0,1]) > 0.$$

   Use this to construct a set E of positive capacity such that every F_σ subset F of E satisfies
$$c(E \setminus F) = c(E).$$

5.2 Computation of Capacity

Though Definition 5.1.1 is fine for the purpose of deriving theoretical properties of capacity, it is not well suited to computing the capacity of specific sets. Even the simplest case, that of a disc, requires some work (see Exercise 5.1.1 above), and most other sets are virtually impossible.

Fortunately, for compact sets at least, there are easier alternatives. They are based on the following relation between capacity and Green's functions.

Theorem 5.2.1 *Let K be a compact non-polar set, and let D be the component of $\mathbb{C}_\infty \setminus K$ which contains ∞. Then*

(5.8) $$g_D(z, \infty) = \log|z| - \log c(K) + o(1) \quad \text{as } z \to \infty.$$

Proof. Let ν be the equilibrium measure for K. From the way that g_D was constructed in Theorem 4.4.2, we have
$$g_D(z, \infty) = p_\nu(z) - I(\nu) = p_\nu(z) - \log c(K) \quad (z \in D \setminus \{\infty\}).$$

By Theorem 3.1.2 we also know that
$$p_\nu(z) = \log|z| + o(1) \quad \text{as } z \to \infty.$$

Combining these two facts gives the result. $\square$

5.2. COMPUTATION OF CAPACITY

As a consequence, we can read off the capacity of a disc.

Corollary 5.2.2 *If $w \in \mathbb{C}$ and $r > 0$, then $c(\overline{\Delta}(w,r)) = r$.*

Proof. Setting $D = \mathbb{C}_\infty \setminus \overline{\Delta}(w,r)$, we have

$$g_D(z,\infty) = \log\left|\frac{z-w}{r}\right| = \log|z| - \log r + o(1) \quad \text{as } z \to \infty.$$

Comparing this with (5.8), we deduce that $c(\overline{\Delta}(w,r)) = r$. □

The subordination principle for Green's functions gives rise to a useful inequality for capacity.

Theorem 5.2.3 *Let K_1 and K_2 be compact subsets of $\mathbb{C}$, and let D_1 and D_2 be the components containing ∞ of $\mathbb{C}_\infty \setminus K_1$ and $\mathbb{C}_\infty \setminus K_2$ respectively. If there is a meromorphic function $f \colon D_1 \to D_2$ such that*

(5.9) $$f(z) = z + O(1) \quad \text{as } z \to \infty,$$

then

$$c(K_2) \leq c(K_1),$$

with equality if f is a conformal mapping of D_1 onto D_2.

Proof. If K_2 is polar, then $c(K_2) = 0$ and the inequality is clear. Thus we may as well suppose that K_2 is non-polar.

Assume, for the moment, that K_1 is also non-polar. Then the Green's functions g_{D_1} and g_{D_2} both exist, and by Theorem 4.4.4

$$g_{D_2}(f(z),\infty) \geq g_{D_1}(z,\infty) \quad (z \in D_1).$$

Now from Theorem 5.2.1, as $z \to \infty$,

$$g_{D_1}(z,\infty) = \log|z| - \log c(K_1) + o(1),$$

and from (5.9),

$$\begin{aligned}g_{D_2}(f(z),\infty) &= \log|f(z)| - \log c(K_2) + o(1)\\ &= \log|z| - \log c(K_2) + o(1).\end{aligned}$$

Combining these facts, we deduce that $c(K_2) \leq c(K_1)$ in this case.

For a general K_1, take $\epsilon > 0$ and set

$$K_1^\epsilon = \{z : \operatorname{dist}(z, K_1) \leq \epsilon\}.$$

This set is non-polar, so by the case just proved we have $c(K_2) \leq c(K_1^\epsilon)$. Letting $\epsilon \to 0$ and applying Theorem 5.1.3 (a), we again obtain $c(K_2) \leq c(K_1)$ (and so, in fact, K_1 was non-polar anyway).

Finally, if f is a conformal mapping of D_1 onto D_2, then we can apply the same argument to f^{-1} to deduce that $c(K_1) \leq c(K_2)$, and hence that equality holds. □

Using this, we can find the capacity of an interval.

Corollary 5.2.4 *If $a \le b$, then $c([a,b]) = (b-a)/4$.*

Proof. The function $f(z) = z + 1/z$ maps $\mathbb{C}_\infty \setminus \overline{\Delta}(0,1)$ conformally onto $\mathbb{C}_\infty \setminus [-2,2]$ and satisfies (5.9), so by Theorem 5.2.3
$$c([-2,2]) = c(\overline{\Delta}(0,1)) = 1.$$
For general a,b, the result follows by translating and scaling. □

In principle, the same technique works for any compact connected set K with more than one point. For by the Riemann mapping theorem, $\mathbb{C}_\infty \setminus K$ can be mapped conformally onto the unit disc, and, by composing with a suitable Möbius transformation, we can find $r > 0$, and a conformal map $f \colon \mathbb{C}_\infty \setminus K \to \mathbb{C}_\infty \setminus \overline{\Delta}(0,r)$ which satisfies (5.9). The capacity of K is then given by
$$c(K) = c(\overline{\Delta}(0,r)) = r.$$
In practice, however, it is only possible to compute the conformal map f explicitly for relatively simple sets K, such as those bounded by a finite number of straight lines and circular arcs. The results of several such calculations are listed in Table 5.1, and details of some of the easier ones are outlined in the exercises.

Capacity also behaves well under taking inverse images by polynomials.

Theorem 5.2.5 *Let K be a compact set, and let $q(z) = \sum_{j=0}^{d} a_j z^j$, where $a_d \ne 0$. Then*
$$c(q^{-1}(K)) = \left(\frac{c(K)}{|a_d|}\right)^{1/d}.$$

Proof. Let D and D' be the components containing ∞ of $\mathbb{C}_\infty \setminus K$ and $\mathbb{C}_\infty \setminus q^{-1}(K)$ respectively. Then, as is easily checked, $q(D') = D$ and $q(\partial D') = \partial D$.

Assume, for the moment, that D is a regular domain. Then by Theorem 4.4.9,
$$\lim_{\substack{z \to \zeta \\ z \in D'}} g_D(q(z), \infty) = 0 \quad (\zeta \in \partial D').$$
Also, $g_D(q(z), \infty)$ is harmonic on $D' \setminus \{\infty\}$, and as $z \to \infty$,
$$g_D(q(z), \infty) = \log |q(z)| + O(1) = d \log |z| + O(1).$$
By the uniqueness of Green's functions, it follows that
$$g_D(q(z), \infty) = d\, g_{D'}(z, \infty) \quad (z \in D').$$

5.2. COMPUTATION OF CAPACITY

Table 5.1: Examples of Capacities

K	$c(K)$
disc, radius r	r
line segment, length h	$h/4$
ellipse, semi-axes a, b	$\dfrac{a+b}{2}$
equilateral triangle, side h	$\dfrac{3^{1/2}\Gamma(1/3)^3}{8\pi^2}h \approx 0.42175h$
isosceles right triangle, short side h	$\dfrac{3^{3/4}\Gamma(1/4)^2}{2^{7/2}\pi^{3/2}}h \approx 0.47563h$
square, side h	$\dfrac{\Gamma(1/4)^2}{4\pi^{3/2}}h \approx 0.59017h$
regular n-gon, side h	$\dfrac{\Gamma(1/n)}{2^{1+2/n}\pi^{1/2}\Gamma(1/2+1/n)}h$
rhombus, side h, angle α	$\dfrac{\pi^{1/2}}{2\Gamma(1-\alpha/2\pi)\Gamma(1/2+\alpha/2\pi)}h$
circular arc, radius r, angle α	$r\sin(\alpha/4)$
half-disc, radius r	$\dfrac{4}{3^{3/2}}r \approx 0.76980r$
lune between circular arcs at angles $\alpha \leq \beta$ to chord of length h	$\dfrac{h}{2\gamma\sin((\pi-\beta)/\gamma)}, \ \gamma = 2+\dfrac{\alpha-\beta}{\pi}$
lemniscate $\{z : \|a_d z^d + \cdots + a_0\| \leq r\}$	$\left(\dfrac{r}{\|a_d\|}\right)^{1/d}$
star $\overline{\Delta}(0,r) \cup \bigcup_{k=1}^{n}[0, Re^{2\pi ik/n}] \ (r < R)$	$r\left(\dfrac{(R/r)^n + 2 + (r/R)^n}{4}\right)^{1/n}$

From Theorem 5.2.1 we also know that, as $z \to \infty$,

$$\begin{aligned} g_D(q(z), \infty) &= \log|q(z)| - \log c(K) + o(1) \\ &= d\log|z| + \log|a_d| - \log c(K) + o(1) \end{aligned}$$

and

$$g_{D'}(z, \infty) = \log|z| - \log c(q^{-1}(K)) + o(1).$$

Putting these facts together, we obtain

$$d\log c\left(q^{-1}(K)\right) = \log c(K) - \log|a_d|,$$

which gives the result in this case.

For a general K, take $\epsilon > 0$ and set

$$K^\epsilon = \{z : \operatorname{dist}(z, K) \leq \epsilon\}.$$

Since no component of K^ϵ is a singleton, it follows that the corresponding domain D_ϵ is regular, and so, by what we have already proved,

$$c\left(q^{-1}(K^\epsilon)\right) = \left(\frac{c(K^\epsilon)}{|a_d|}\right)^{1/d}.$$

The result follows by letting $\epsilon \to 0$ and applying Theorem 5.1.3 (a). □

This result can be used to compute the capacity of a few disconnected sets which possess some symmetry. As an illustration, we do this for a union of two intervals of equal length.

Corollary 5.2.6 *If $0 \leq a \leq b$, then $c\left([-b, -a] \cup [a, b]\right) = \sqrt{b^2 - a^2}/2$.*

Proof. Taking $q(z) = z^2$, we have

$$c\left([-b, -a] \cup [a, b]\right) = c\left(q^{-1}[a^2, b^2]\right) = c\left([a^2, b^2]\right)^{1/2} = \left((b^2 - a^2)/4\right)^{1/2},$$

whence the result. □

Exercises 5.2

1. Show that if $q(z) = z^2 - 2$, then $q^{-1}\left([-2, 2]\right) = [-2, 2]$. Hence give another proof that $c([-2, 2]) = 1$.

2. Show that if $r \geq 1$, then $f(z) := z + 1/z$ maps $\mathbb{C}_\infty \setminus \overline{\Delta}(0,r)$ conformally onto $\mathbb{C}_\infty \setminus K$, where K is an ellipse with semi-axes $a := r + 1/r$ and $b := r - 1/r$. Deduce that

$$c(K) = \frac{a+b}{2}.$$

3. Let $K = \overline{\Delta}(0,1) \cup [0, R]$, where $R \geq 1$. Show that $f(z) := z + 1/z$ maps $\mathbb{C}_\infty \setminus K$ conformally onto $\mathbb{C}_\infty \setminus [-2, R + 1/R]$, and deduce that

$$c(K) = \frac{R + 2 + 1/R}{4}.$$

By considering $q^{-1}(K)$, where $q(z) = z^n$, derive the last entry in Table 5.1.

4. Let K be the circular arc $\{e^{i\theta} : |\theta| \leq \alpha/2\}$, where $0 < \alpha < 2\pi$, and let

$$f(z) = \frac{1}{2}\left(z - 1 + \sqrt{(z - e^{i\alpha/2})(z - e^{-i\alpha/2})}\right),$$

where the square root is chosen so that $f(z) = z + O(1)$ as $z \to \infty$. Show that f maps $\mathbb{C}_\infty \setminus K$ conformally onto $\mathbb{C}_\infty \setminus \overline{\Delta}(0, \sin(\alpha/4))$, and deduce that

$$c(K) = \sin(\alpha/4).$$

5.3 Estimation of Capacity

Even for relatively simple sets, such as a square, calculation of the capacity requires some effort. For more complicated sets it is usually impossible, and we have to be content with estimates. In this section we shall derive various upper and lower bounds for capacity in terms of other, more easily computed geometric quantities. As in the previous section, we shall restrict attention to compact sets, relying on results such as Theorem 5.1.2 (b) to cater for more general sets.

Many of the estimates rely on the following basic result.

Theorem 5.3.1 *Let K be a compact subset of $\mathbb{C}$, and let $T: K \to \mathbb{C}$ be a map satisfying*

(5.10) $\qquad |T(z) - T(w)| \leq A|z - w|^\alpha \quad (z, w \in K),$

where A and α are positive constants. Then

$$c(T(K)) \leq A\, c(K)^\alpha.$$

Proof. Let ν be an equilibrium measure for the compact set $T(K)$. By Theorem A.4.4 from the Appendix, there exists a Borel probability measure μ on K such that $\mu T^{-1} = \nu$. Then

$$\begin{aligned} I(\nu) &= \int_K \int_K \log|T(z) - T(w)|\, d\mu(z)\, d\mu(w) \\ &\leq \int_K \int_K \log(A|z-w|^\alpha)\, d\mu(z)\, d\mu(w) \\ &= \log A + \alpha I(\mu). \end{aligned}$$

Hence, from the definition of capacity, we have

$$c(T(K)) = e^{I(\nu)} \leq A e^{\alpha I(\mu)} \leq A\, c(K)^\alpha,$$

which proves the result. □

Using this theorem in conjunction with Corollary 5.2.4, we deduce a number of '$\frac{1}{4}$-estimates' for capacity.

Theorem 5.3.2 *Let K be a compact subset of $\mathbb{C}$.*

(a) *If K is connected and has diameter d, then*

$$c(K) \geq d/4.$$

(b) *If K is a rectifiable curve of length l, then*

$$c(K) \leq l/4.$$

(c) *If K is a subset of the real axis of Lebesgue measure m, then*

$$c(K) \geq m/4.$$

(d) *If K is a subset of the unit circle of arc-length measure a, then*

$$c(K) \geq \sin(a/4).$$

The example of a line segment (or that of a circular arc in case (d)) shows that all these inequalities are sharp.

Proof. (a) Rotating and translating, we can suppose that $0, d \in K$. Let $T: \mathbb{C} \to \mathbb{R}$ denote the orthogonal projection onto the real axis. Then $T(K)$ is a connected set containing $0, d$, so it contains $[0, d]$, and hence

$$c(T(K)) \geq c([0, d]) = d/4.$$

On the other hand, T satisfies (5.10) with $A = \alpha = 1$, so by Theorem 5.3.1

$$c(T(K)) \leq c(K).$$

The result follows.

5.3. ESTIMATION OF CAPACITY

(b) Let $T\colon [0,l] \to K$ be the arc-length parametrization of K. Then T satisfies (5.10) with $A = \alpha = 1$, so by Theorem 5.3.1
$$c(K) \leq c([0,l]) = l/4.$$

(c) Define $T\colon \mathbb{R} \to \mathbb{R}$ by $T(x) = $ Lebesgue measure of $(K \cap (-\infty, x])$. Then $T(K) = [0,m]$, so
$$c(T(K)) = c([0,m]) = m/4.$$
Again T satisfies (5.10) with $A = \alpha = 1$, so the result follows as in (a).

(d) If K is contained within a semicircle, then one can employ an argument similar to that in (c), using the result of Exercise 5.2.4 in place of Corollary 5.2.4. For general K, however, it necessary to proceed somewhat differently.

Define $f_1 \colon \mathbb{C}_\infty \setminus K \to \mathbb{C}$ by
$$f_1(z) = \frac{1}{4} \int_K \frac{z+\zeta}{z-\zeta} |d\zeta|,$$
so that f_1 is holomorphic on $\mathbb{C}_\infty \setminus K$, with $f_1(\infty) = a/4$ and $f_1(0) = -a/4$. Also
$$\operatorname{Re} f_1(z) = \frac{1}{4} \int_K \frac{|z|^2 - 1}{|z-\zeta|^2} |d\zeta| = -\frac{\pi}{2} \int_K P(z,\zeta)\, |d\zeta|,$$
from which it follows that

(5.11) $\qquad -\pi/2 \leq \operatorname{Re} f_1(z) \leq \pi/2 \quad (z \in \mathbb{C}_\infty \setminus K).$

Now define $f_2 \colon \mathbb{C}_\infty \setminus K \to \mathbb{C}$ by
$$f_2(z) = \frac{e^{if_1(z)} - e^{-ia/4}}{e^{if_1(z)} + e^{ia/4}},$$
so that f_2 is holomorphic on $\mathbb{C}_\infty \setminus K$, with $f_2(\infty) = ie^{-ia/4}\sin(a/4)$ and $f_2(0) = 0$. Also (5.11) implies that $|f_2(z)| \leq 1$ for all $z \in \mathbb{C}_\infty \setminus K$, and so, using Schwarz's lemma, it follows that
$$\left|\frac{f_2(z)}{z}\right| < 1 \quad (z \in \mathbb{C}_\infty \setminus K).$$

Finally, define $f_3 \colon \mathbb{C}_\infty \setminus K \to \mathbb{C}_\infty$ by
$$f_3(z) = f_2(\infty) \frac{z}{f_2(z)},$$
so that f_3 is meromorphic on $\mathbb{C}_\infty \setminus K$, with $f_3(z) = z + O(1)$ as $z \to \infty$. Then
$$|f_3(z)| > |f_2(\infty)| = \sin(a/4) \quad (z \in \mathbb{C}_\infty \setminus K),$$
and so from Theorem 5.2.3 we deduce that $c(K) \geq \sin(a/4)$, as desired. $\square$

As an application of this result, we prove the celebrated Koebe one-quarter theorem for univalent functions.

Theorem 5.3.3 (Koebe's One-Quarter Theorem) *If f is an injective holomorphic function on $\Delta(0,1)$ with $f(0) = 0$ and $f'(0) = 1$, then*

$$f(\Delta(0,1)) \supset \Delta(0, 1/4).$$

That $f(\Delta(0,1))$ contains *some* disc about the origin is a consequence of the open mapping theorem. The point of Koebe's theorem is that this disc always has radius at least $1/4$. The constant $1/4$ is sharp, as can be seen by considering the function $f(z) = z/(1-z)^2$.

Proof. Let K be the compact set given by

$$K = \{z \in \mathbb{C} : 1/z \notin f(\Delta(0,1))\},$$

and define $f_1 \colon \mathbb{C}_\infty \setminus \overline{\Delta}(0,1) \to \mathbb{C}_\infty \setminus K$ by

$$f_1(z) = \frac{1}{f(1/z)}.$$

Then f_1 is a conformal homeomorphism, and $f_1(z) = z + O(1)$ as $z \to \infty$, so by Theorem 5.2.3

$$c(K) = c(\overline{\Delta}(0,1)) = 1.$$

Also $\mathbb{C}_\infty \setminus K$ is homeomorphic to $\mathbb{C}_\infty \setminus \overline{\Delta}(0,1)$, which is simply connected, and hence K is connected. Therefore by Theorem 5.3.2 (a),

$$\mathrm{diam}(K) \leq 4c(K) = 4.$$

As $0 \in K$, we deduce that $K \subset \overline{\Delta}(0,4)$, from which the result follows. □

As we saw in Theorem 5.1.4, it is an easy consequence of the definition of capacity that $c(K) \leq \mathrm{diam}(K)$ for every compact set K. But in fact this can be improved.

Theorem 5.3.4 *If K is a compact subset of $\mathbb{C}$ of diameter d, then*

$$c(K) \leq d/2.$$

The example of a disc shows that this inequality *is* sharp.

5.3. ESTIMATION OF CAPACITY

Proof. Replacing K by its convex hull, which increases the capacity but leaves the diameter unchanged, we can assume that K is convex. We may also suppose that K contains more than one point, so by the Riemann mapping theorem there exists a conformal map $f\colon \mathbb{C}_\infty \setminus K \to \mathbb{C}_\infty \setminus \overline{\Delta}(0,1)$ with $f(\infty) = \infty$. Define $u\colon \mathbb{C} \setminus K \to [-\infty, \infty)$ by

$$u(z) = \log\left|\frac{z - f^{-1}(-f(z))}{d}\right| - g_{\mathbb{C}_\infty \setminus K}(z, \infty),$$

so that u is subharmonic on $\mathbb{C} \setminus K$. Then, using Theorem 5.2.1, we have

$$u(z) = \log\left|\frac{2z}{d}\right| - \log|z| + \log c(K) + o(1) \quad \text{as } z \to \infty,$$

and so we can remove the singularity at ∞ by setting

$$u(\infty) = \log\left(\frac{2}{d}\right) + \log c(K).$$

Now $\operatorname{dist}\bigl(f^{-1}(-f(z)), \partial K\bigr) \to 0$ as $\operatorname{dist}(z, \partial K) \to 0$, and so

$$\limsup_{z \to \zeta} u(z) \le \log\left|\frac{d}{d}\right| - 0 = 0 \quad (\zeta \in \partial K).$$

Hence by the maximum principle $u \le 0$ on $\mathbb{C}_\infty \setminus K$, and in particular $u(\infty) \le 0$, which gives the result. $\square$

Since there are sets, such as line segments, which have positive capacity but zero area, we would not expect to find an upper bound for capacity in terms of area. But there is a lower bound, which can be viewed as a kind of isoperimetric inequality for capacity.

Theorem 5.3.5 *If K is a compact subset of $\mathbb{C}$ of area A, then*

$$c(K) \ge \sqrt{A/\pi}.$$

The example of a disc shows that this inequality is sharp, though if K is connected then it can be generalized to take account of the 'dispersion' of K (see Exercise 5). The proof of Theorem 5.3.5 proceeds via a lemma which is of interest in its own right.

Lemma 5.3.6 (Ahlfors–Beurling Inequality) *If K is a compact subset of $\mathbb{C}$ of area A, then*

$$\left|\int_K \frac{1}{w-z}\, dA(w)\right| \le \sqrt{\pi A} \quad (z \in \mathbb{C}).$$

Proof. We begin by making some reductions. First of all, if K has zero area then the inequality is obvious, so we may as well assume that $A > 0$. Also, it is enough to prove the inequality for the special case $z = 0$; the general case then follows by applying this to the translate $K - z$. Finally, we can suppose that $\int_K w^{-1} dA(w) \geq 0$, otherwise just rotate K about the origin until it becomes true.

Let Δ be the disc $\{w : \text{Re}(1/w) > 1/(2a)\}$, where the radius a is chosen so that Δ and K have equal area, in other words $\pi a^2 = A$. Then

$$\begin{aligned}
\int_K \frac{1}{w} dA(w) &= \int_K \text{Re}\left(\frac{1}{w}\right) dA(w) \\
&\leq \int_{K \cap \Delta} \text{Re}\left(\frac{1}{w}\right) dA(w) + \int_{K \setminus \Delta} \frac{1}{2a} dA(w) \\
&= \int_{K \cap \Delta} \text{Re}\left(\frac{1}{w}\right) dA(w) + \int_{\Delta \setminus K} \frac{1}{2a} dA(w) \\
&\leq \int_\Delta \text{Re}\left(\frac{1}{w}\right) dA(w) \\
&= \int_{-\pi/2}^{\pi/2} \int_0^{2a\cos\theta} \frac{\cos\theta}{r} r \, dr \, d\theta \\
&= \pi a = \sqrt{\pi A}.
\end{aligned}$$

This is the desired inequality. $\square$

Proof of Theorem 5.3.5. Let D be the component of $\mathbb{C}_\infty \setminus K$ containing ∞, and define $f: D \to \mathbb{C}_\infty$ by

$$f(z) = \left(\frac{1}{A} \int_K \frac{1}{z - w} dA(w)\right)^{-1}.$$

Then f is meromorphic, $f(z) = z + O(1)$ as $z \to \infty$, and by Lemma 5.3.6 f maps D into $\mathbb{C}_\infty \setminus \overline{\Delta}(0, \sqrt{A/\pi})$. Hence by Theorem 5.2.3,

$$c(K) \geq c\left(\overline{\Delta}(0, \sqrt{A/\pi})\right) = \sqrt{A/\pi},$$

as claimed. $\square$

Finally, in this section, we return to the problem mentioned at the beginning of the chapter, to determine whether or not the Cantor set is polar. In fact we shall study generalized Cantor sets, constructed as follows.

Let $\mathbf{s} := (s_n)_{n \geq 1}$ be a sequence of numbers such that $0 < s_n < 1$ for all n. Define $C(s_1)$ to be the set obtained from $[0,1]$ by removing an open interval

5.3. ESTIMATION OF CAPACITY

of length s_1 from the centre. At the n-th stage, let $C(s_1, \ldots, s_n)$ be the set obtained by removing from the middle of each interval in $C(s_1, \ldots, s_{n-1})$ an open subinterval whose length is a proportion s_n of the whole interval. We thereby obtain a decreasing sequence of compact sets $(C(s_1, \ldots, s_n))_{n \geq 1}$, and the corresponding *generalized Cantor set* is defined to be

$$C(\mathbf{s}) := \bigcap_{n \geq 1} C(s_1, \ldots, s_n).$$

It is readily checked that $C(\mathbf{s})$ is a compact, perfect, totally disconnected set of Lebesgue measure $\prod_{n \geq 1}(1 - s_n)$. We now investigate its capacity.

Theorem 5.3.7 *With the notation above,*

$$\frac{pq}{2} \leq c(C(\mathbf{s})) \leq \frac{p}{2},$$

where $p = \prod_{n \geq 1}(1 - s_n)^{1/2^n}$ *and* $q = \prod_{n \geq 1} s_n^{1/2^n}$.

Thus, for example, the standard Cantor set, which is obtained by taking $s_n = 1/3$ for all n, has capacity at least $1/9$, and in particular it is non-polar. On the other hand, if we let $s_n = 1 - (1/2)^{2^n}$, then $C(\mathbf{s})$ is polar, thereby providing the long-promised example of an uncountable polar set.

Proof. We begin by proving the upper bound. Put $K = C(s_1, \ldots, s_n)$, and let K_1, K_2 denote the left-hand and right-hand halves of K respectively. As $\operatorname{diam}(K) = 1$, we can apply Theorem 5.1.4 (a) with $d = 1$ to obtain

$$\frac{1}{\log(1/c(K))} \leq \sum_{j=1}^{2} \frac{1}{\log(1/c(K_j))}.$$

By symmetry $c(K_1) = c(K_2)$, so the above inequality simplifies to

$$\log c(K) \leq \frac{1}{2} \log c(K_1).$$

Now $K = C(s_1, \ldots, s_n)$, and K_1 is just the set $C(s_2, \ldots, s_n)$ scaled down by a factor $(1 - s_1)/2$, so the inequality becomes

$$\log(c(C(s_1, \ldots, s_n))) \leq \frac{1}{2} \log\left(\frac{1 - s_1}{2}\right) + \frac{1}{2} \log c(C(s_2, \ldots, s_n)).$$

Iterating this gives

$$\log(c(C(s_1, \ldots, s_n))) \leq \sum_{j=1}^{n} \frac{1}{2^j} \log\left(\frac{1 - s_j}{2}\right) + \frac{1}{2^n} \log c([0, 1]).$$

Letting $n \to \infty$, we obtain

$$\log c(C(\mathbf{s})) \leq \sum_{j=1}^{\infty} \frac{1}{2^j} \log\left(\frac{1-s_j}{2}\right),$$

which gives the upper bound.

The lower bound is proved in a similar fashion. With K, K_1, K_2 as before, we have $\mathrm{dist}(K_1, K_2) = s_1$, so applying Theorem 5.1.4 (b) with $d = s_1$ gives

$$\frac{1}{\log^+(s_1/c(K))} \geq \sum_{j=1}^{2} \frac{1}{\log^+(s_1/c(K_j))}.$$

If $c(K) < s_1$, then this simplifies to

$$\log c(K) \geq \frac{1}{2} \log s_1 + \frac{1}{2} \log c(K_1),$$

and if $c(K) \geq s_1$, then this inequality is clear anyway, since $c(K) \geq c(K_1)$. Repeating the argument used for the upper bound now leads to

$$\log c(C(\mathbf{s})) \geq \sum_{j=1}^{\infty} \frac{1}{2^j} \log s_j + \sum_{j=1}^{\infty} \frac{1}{2^j} \log\left(\frac{1-s_j}{2}\right),$$

which gives the lower bound. □

Exercises 5.3

1. (i) Show that Theorem 5.3.1 remains true for F_σ sets.

    (ii) Assuming Choquet's theorem (5.3), deduce that Theorem 5.3.1 holds for arbitrary Borel sets. Hence show that every Borel polar set is totally disconnected.

    (iii) Let $\mathbb{C} = P \cup Q$ be a partition of $\mathbb{C}$ into subsets P, Q, neither of which contains an uncountable compact set (see Exercise 3.2.2). Show that P and Q are necessarily connected, and deduce that non-Borel polar sets need not be totally disconnected.

    [Hint: Suppose there exist closed subsets F_1, F_2 of $\mathbb{C}$ such that $P \subset F_1 \cup F_2$ and $P \cap F_j \neq \emptyset$ ($j=1,2$), but $P \cap F_1 \cap F_2 = \emptyset$. Pick points $z_j \in P \cap F_j$. Show that if C is any curve joining z_1 and z_2, then $C \subset F_1 \cup F_2$ and $C \cap F_1 \cap F_2 \neq \emptyset$. Deduce that $F_1 \cap F_2$ is uncountable, and hence derive a contradiction.]

5.3. ESTIMATION OF CAPACITY

2. Let K be a compact connected set that contains both $\overline{\Delta}(0,1)$ and the point $R > 1$. Show that
$$c(K) \geq \frac{R+2+1/R}{4}.$$

3. Let K be a compact subset of $\mathbb{C}$.

   (i) Show that $c(K) \geq m/4$, where m is the Lebesgue measure of the set
   $$\{r \in [0,\infty) : re^{i\theta} \in K \text{ for some } \theta\}.$$

   (ii) Show that if $\rho > 0$ then $c(K) \geq \rho \sin(a/4)$, where a is the Lebesgue measure of the set
   $$\{\theta \in [0, 2\pi) : re^{i\theta} \in K \text{ for some } r \geq \rho\}.$$

4. Use the result of the previous exercise to prove the following generalization of the Koebe one-quarter theorem: if $f \colon \Delta(0,1) \to \mathbb{C}$ is holomorphic with $f(0) = 0$ and $f'(0) = 1$, then the set
$$\{r \in [0,\infty) : \partial \Delta(0,r) \subset f(\Delta(0,1))\}$$
has Lebesgue measure at least $1/4$.

5. Let K be a connected compact set, and let $n \geq 1$.

   (i) Let f be a conformal map of $\mathbb{C}_\infty \setminus \overline{\Delta}(0, c(K))$ onto the component of $\mathbb{C}_\infty \setminus K$ which contains ∞, such that $f(z) = z + O(1)$ as $z \to \infty$. Show that there is a monic polynomial q_0 of degree n such that
   $$q_0 \circ f(z) = z^n + O(1) \quad \text{as } z \to \infty.$$

   (ii) Given $r > c(K)$, let K_r be the compact set defined by $\mathbb{C}_\infty \setminus K_r = f(\mathbb{C}_\infty \setminus \overline{\Delta}(0,r))$. Use Green's theorem to show that
   $$\int_{K_r} |q_0'(z)|^2 \, dA(z) = \frac{1}{2i} \int_{\partial K_r} \overline{q_0(z)} q_0'(z) \, dz \leq n\pi r^{2n}.$$

   (iii) Conclude that
   $$c(K) \geq \inf_q \left(\frac{1}{n\pi} \int_K |q'(z)|^2 \, dA(z) \right)^{1/2n},$$
   where the inf is taken over all monic polynomials q of degree n.

(iv) By taking $n = 1, 2$, deduce that
$$c(K) \geq (A/\pi)^{1/2} \quad \text{and} \quad c(K) \geq (2I/\pi)^{1/4},$$
where A is the area of K, and I is the moment of inertia about its centre of mass. Explain why the second inequality is stronger than the first.

6. Let γ be a closed piecewise-smooth path in $\mathbb{C}$, and for each integer k let
$$U_k = \{z \in (\mathbb{C} \setminus \gamma) : \gamma \text{ winds } k \text{ times around } z\}.$$
Show that if $K = (\cup_{k \geq 1} U_k) \cup \gamma$, then
$$\int_K \int_\gamma \frac{1}{z-w} \, dz \, dA(w) = 2\pi i \sum_{k \geq 1} k \, \text{area}(U_k).$$
Use Lemma 5.3.6 to deduce the following form of the isoperimetric inequality:
$$\sum_{k \geq 1} k \, \text{area}(U_k) \leq \frac{\text{length}(\gamma)^2}{4\pi}.$$

7. Let Δ be the unit disc. Give an example of a subset of $\partial \Delta$ which is non-polar, but which has harmonic measure zero for Δ. [Hint: consider an image of the Cantor set.]

5.4 Criteria for Thinness

As we saw in Theorem 4.2.4, the question of whether a given point is regular for the Dirichlet problem on a domain D is equivalent to whether $\mathbb{C}_\infty \setminus D$ is non-thin at the point. Unfortunately, at that time we had no general criterion for thinness, but with the theory of capacity at our disposal, we are now in a position to put that right.

Theorem 5.4.1 (Wiener's Criterion) *Let F be an F_σ subset of $\mathbb{C}$, and let $\zeta_0 \in \mathbb{C}$. Let γ be a constant with $0 < \gamma < 1$, and for $n \geq 1$ define*
$$F_n = \{z \in F : \gamma^n < |z - \zeta_0| \leq \gamma^{n-1}\}.$$
Then F is thin at ζ_0 if and only if

(5.12) $$\sum_{n \geq 1} \frac{n}{\log(2/c(F_n))} < \infty.$$

5.4. CRITERIA FOR THINNESS

Proof. Since thinness and capacity both remain invariant under translation, we may as well suppose from the outset that $\zeta_0 = 0$. We can also assume that $0 \notin F$, and that $F_n \neq \emptyset$ for each n (otherwise just remove 0, and add an appropriate countable set).

Assume first that (5.12) holds: we shall show that F is thin at 0. As each F_n is an F_σ set, we may write it as $F_n = \cup_m K_{nm}$, where $(K_{nm})_{m \geq 1}$ is an increasing sequence of compact sets. For each pair n, m, let ν_{nm} be an equilibrium measure for K_{nm}. Then by Frostman's theorem

$$p_{\nu_{nm}} = I(\nu_{nm}) = \log c(K_{nm}) \leq \log c(F_n) \quad \text{n.e. on } K_{nm}.$$

Also, as $K_{nm} \subset \overline{\Delta}(0,1)$, which has diameter 2, we have

$$p_{\nu_{nm}} \leq \log 2 \quad \text{on } \overline{\Delta}(0,1).$$

Lastly, since $K_{nm} \cap \overline{\Delta}(0, \gamma^n) = \emptyset$, it follows that

$$p_{\nu_{nm}}(0) = \int_{K_{nm}} \log |w| \, d\nu_{nm}(w) \geq n \log \gamma.$$

Now set $\alpha_n = 1/\log(2/c(F_n))$. By our assumption, $\sum_n n\alpha_n < \infty$, so we can find a sequence of positive numbers (β_n) such that $\beta_n \to \infty$ and still $\sum_n n\alpha_n\beta_n < \infty$. For each $m \geq 1$, define u_m on $\Delta(0,1)$ by

$$u_m = \sum_{n \geq 1} \alpha_n \beta_n (p_{\nu_{nm}} - \log 2).$$

Then by Theorem 2.4.6 u_m is subharmonic on $\Delta(0,1)$, and

$$\begin{aligned} u_m &\leq -\beta_n \quad \text{n.e. on } K_{nm}, \\ u_m &\leq 0 \quad \text{on } \Delta(0,1), \\ u_m(0) &\geq \sum_{n \geq 1} \alpha_n \beta_n (n \log \gamma - \log 2). \end{aligned}$$

Next, define u on $\Delta(0,1)$ by

$$u = \left(\limsup_{m \to \infty} u_m\right)^*.$$

Then by Theorem 3.4.3 u is subharmonic on $\Delta(0,1)$, and

$$\begin{aligned} u &\leq -\beta_n \quad \text{n.e. on } F_n, \\ u &\leq 0 \quad \text{on } \Delta(0,1), \\ u(0) &\geq \sum_{n \geq 1} \alpha_n \beta_n (n \log \gamma - \log 2). \end{aligned}$$

In particular, if we set
$$E = \bigcup_{n \geq 1} \{z \in F_n : u(z) > -\beta_n\},$$
then
$$\limsup_{\substack{z \to 0 \\ z \in F \setminus E}} u(z) \leq \lim_{n \to \infty} -\beta_n = -\infty < u(0).$$

Therefore $F \setminus E$ is thin at 0. But E is an F_σ polar set, so by Theorem 3.8.2 it too is thin at 0. Hence F is thin at 0, as was to be shown.

Now assume that F is thin at 0: we shall prove that (5.12) holds. By our assumption, there exists a function u, subharmonic on a neighbourhood of 0, such that
$$\limsup_{\substack{z \to 0 \\ z \in F}} u(z) < u(0).$$

By the Riesz decomposition theorem, we may take u to be of the form $u = p_\mu$, where μ is a finite Borel measure on $\overline{\Delta}(0, 1)$. Then in particular $p_\mu(0) > -\infty$, and hence, writing $A_k = \{w : \gamma^k < |w| \leq \gamma^{k-1}\}$, we have
$$p_\mu(z) - p_\mu(0) = \sum_{k \geq 1} \int_{A_k} \log\left|1 - \frac{z}{w}\right| d\mu(w) \quad (z \in \mathbb{C}).$$

Now if $z \in A_n$ and $w \in A_k$, where $k \leq n - 2$, then $|z/w| \leq \gamma^{n-k-1}$, and so
$$\inf_{z \in A_n} \sum_{k=1}^{n-2} \int_{A_k} \log\left|1 - \frac{z}{w}\right| d\mu(w) \geq \sum_{k=1}^{n-2} \log(1 - \gamma^{n-k-1}) \mu(A_k)$$
$$\to 0 \quad \text{as } n \to \infty.$$

Also if $z \in A_n$ and $w \in A_k$, where $k \geq n + 2$, then $|z/w| \geq \gamma^{-1}$, and so
$$\inf_{z \in A_n} \sum_{k=n+2}^{\infty} \int_{A_k} \log\left|1 - \frac{z}{w}\right| d\mu(w) \geq \sum_{k=n+2}^{\infty} \log(\gamma^{-1} - 1) \mu(A_k)$$
$$\to 0 \quad \text{as } n \to \infty.$$

Lastly, since μ is supported on $\overline{\Delta}(0, 1)$, we have
$$\sum_{k=n-1}^{n+1} \int_{A_k} \log|w| \, d\mu(w) \leq 0.$$

5.4. CRITERIA FOR THINNESS

It follows that, given $\epsilon > 0$, there exists n_0 such that, for all $n \geq n_0$,

$$\sum_{k=n-1}^{n+1} \int_{A_k} \log|z-w|\,d\mu(w) \leq p_\mu(z) - p_\mu(0) + \epsilon \quad (z \in A_n).$$

We shall apply this with ϵ chosen small enough so that

$$\limsup_{\substack{z \to 0 \\ z \in F}} p_\mu(z) < p_\mu(0) - 2\epsilon.$$

Then, increasing n_0 if necessary, we have that for all $n \geq n_0$,

$$\sum_{k=n-1}^{n+1} \int_{A_k} \log|z-w|\,d\mu(w) \leq -\epsilon \quad (z \in F_n).$$

For each $n \geq n_0$, write $F_n = \cup_m K_{nm}$, where $(K_{nm})_{m \geq 1}$ is an increasing sequence of compact sets, and let ν_{nm} be an equilibrium measure for K_{nm}. Then integrating the last inequality with respect to ν_{nm} yields

$$\sum_{k=n-1}^{n+1} \int_{A_k} p_{\nu_{nm}}(w)\,d\mu(w) \leq -\epsilon.$$

Now by Frostman's theorem

$$p_{\nu_{nm}} \geq I(\nu_{nm}) = \log c(K_{nm}) \quad \text{on } \mathbb{C}.$$

Hence

$$\sum_{k=n-1}^{n+1} \log c(K_{nm})\mu(A_k) \leq -\epsilon \quad (n \geq n_0,\ m \geq 1).$$

Letting $m \to \infty$ and rearranging, we obtain

$$\frac{1}{\log(1/c(F_n))} \leq \frac{1}{\epsilon} \sum_{k=n-1}^{n+1} \mu(A_k) \quad (n \geq n_0).$$

Thus to prove that (5.12) holds, it is enough to show that

$$\sum_{n \geq 1}(n-1)\mu(A_n) < \infty.$$

This is done by observing that if $w \in A_n$, then $\log|w| \leq -(n-1)\log(1/\gamma)$, and therefore

$$\sum_{n \geq 1}(n-1)\mu(A_n) \leq -\sum_{n \geq 1}\int_{A_n}\frac{\log|w|}{\log(1/\gamma)}\,d\mu(w) = -\frac{p_\mu(0)}{\log(1/\gamma)} < \infty.$$

This completes the proof. $\square$

Even though the criterion (5.12) is rather complicated, it can be combined with the results of the previous section to produce simpler conditions which are necessary for thinness, or equivalently, ones which are sufficient for non-thinness. The first of these is a strengthening of the result of Exercise 3.8.3.

Theorem 5.4.2 *Let F be an F_σ subset of $\mathbb{C}$. If F is thin at 0, then*
$$E := \{r \in (0,1] : re^{i\theta} \in F \text{ for some } \theta\}$$
is a set of finite logarithmic measure, i.e.
$$\int_E \frac{1}{x} \, dx < \infty.$$

Proof. Let $0 < \gamma < 1$, and for $n \geq 1$ define
$$F_n = \{z : \gamma^n < |z| \leq \gamma^{n-1}\}.$$
Let $T : \mathbb{C} \to \mathbb{R}$ denote the circular projection $T(z) = |z|$. Then, by applying Theorems 5.3.2 (c) and 5.3.1 to a sequence of compact sets increasing to F_n, we obtain
$$\int_{T(F_n)} dx \leq 4c(T(F_n)) \leq 4c(F_n).$$
Since $E = \cup_{n \geq 1} T(F_n)$, it follows that
$$\int_E \frac{1}{x} \, dx \leq \sum_{n \geq 1} \int_{T(F_n)} \frac{1}{\gamma^n} \, dx \leq \sum_{n \geq 1} \frac{4c(F_n)}{\gamma^n}.$$
Now $t \leq 1/\log(1/t)$ for $t \in (0,1)$, so
$$\frac{c(F_n)}{\gamma^n} \leq \frac{2}{\log(2\gamma^n/c(F_n))}.$$
Hence
$$\int_E \frac{1}{x} \, dx \leq 8 \sum_{n \geq 1} \frac{1}{\log(2\gamma^n/c(F_n))},$$
and if F is thin at 0, then this last sum is finite by Theorem 5.4.1. □

Using radial projection instead of circular projection leads to a different type of result.

Theorem 5.4.3 *Let F be an F_σ subset of $\mathbb{C}$. If F is thin at 0, then*
$$E := \{e^{i\theta} : r_n e^{i\theta} \in F \text{ for some sequence } r_n \to 0\}$$
is a polar set.

5.4. CRITERIA FOR THINNESS

Proof. Again, let $0 < \gamma < 1$, and define
$$F_n = \{z : \gamma^n < |z| \leq \gamma^{n-1}\}.$$
This time, let $T: \mathbb{C} \setminus \{0\} \to \partial \Delta(0,1)$ be the radial projection $T(z) = z/|z|$. Then, by applying Theorem 5.3.1 to a sequence of compact sets increasing to F_n, we obtain
$$c(T(F_n)) \leq \frac{c(F_n)}{\gamma^n}.$$
Now
$$E = \bigcap_{m \geq 1} \bigcup_{n \geq m} T(F_n),$$
so using Theorem 5.1.4 (a), it follows that for each $m \geq 1$
$$\frac{1}{\log(2/c(E))} \leq \frac{1}{\log(2/c(\cup_{n \geq m} T(F_n)))}$$
$$\leq \sum_{n \geq m} \frac{1}{\log(2/c(T(F_n)))}$$
$$\leq \sum_{n \geq m} \frac{1}{\log(2\gamma^n/c(F_n))}.$$
Again, if F is thin at 0, then Theorem 5.4.1 implies that this last series converges, and hence, letting $m \to \infty$, we deduce that $c(E) = 0$. $\square$

This has the following pleasant consequence.

Corollary 5.4.4 *If u is a function subharmonic on a neighbourhood of 0, then*
$$\lim_{r \to 0} u(re^{i\theta}) = u(0) \quad \text{for n.e. } e^{i\theta}.$$

Proof. For each $k \geq 1$ define
$$U_k = \{z : u(z) < u(0) - 1/k\}.$$
Then U_k is an open set which is thin at 0, and so by Theorem 5.4.3
$$\liminf_{r \to 0} u(re^{i\theta}) \geq u(0) - 1/k \quad \text{for n.e. } e^{i\theta}.$$
As a countable union of Borel polar sets is polar, we deduce that
$$\liminf_{r \to 0} u(re^{i\theta}) \geq u(0) \quad \text{for n.e. } e^{i\theta}.$$
On the other hand, by upper semicontinuity we certainly have
$$\limsup_{r \to 0} u(re^{i\theta}) \leq u(0) \quad \text{for all } e^{i\theta}.$$
Combining these inequalities gives the result. $\square$

Exercises 5.4

1. Let F be an F_σ subset of $\mathbb{C}_\infty$, let $\gamma > 1$, and for $n \geq 1$ define
$$F_n = \{z \in F : \gamma^{n-1} \leq |z| < \gamma^n\}.$$
Use the map $z \mapsto 1/z$ to show that F is thin at ∞ if and only if
$$\sum_{n \geq 1} \frac{n}{\log(2/c(F_n))} < \infty.$$

2. Let F be an F_σ subset of $\mathbb{C}$. Show that if F is thin at ζ_0, then
$$\lim_{r \to 0} \frac{c(F \cap \overline{\Delta}(\zeta_0, r))}{r} = 0.$$
Deduce that the Cantor set is non-thin at each point of itself.

3. Prove the following converse to Corollary 5.4.4: given an F_σ polar subset E of the unit circle, there exists a subharmonic function u on $\mathbb{C}$ such that
$$\liminf_{r \to 0} u(re^{i\theta}) < u(0) \quad (e^{i\theta} \in E).$$
[Hint: choose u so that $u(e^{i\theta}/n) = -\infty$ $(e^{i\theta} \in E, n \geq 1)$.]

4. Assuming the result of Choquet (5.3), show that the Wiener criterion remains valid for arbitrary Borel subsets of $\mathbb{C}$.

5.5 Transfinite Diameter

There is another approach to capacity which is actually more direct than Definition 5.1.1. As well as giving further useful estimates for capacity, it has close links with the theory of uniform approximation. It is based on the following definition.

Definition 5.5.1 Let K be a compact subset of $\mathbb{C}$, and let $n \geq 2$. The *n-th diameter* of K is given by
$$\delta_n(K) := \sup\left\{ \prod_{j,k:j<k} |w_j - w_k|^{2/n(n-1)} : w_1, \ldots, w_n \in K \right\}.$$

An n-tuple $w_1, \ldots, w_n \in K$ for which the supremum is attained is called a *Fekete n-tuple* for K.

As K is compact, a Fekete n-tuple always exists, though it need not be unique. The maximum principle shows that in fact it must lie in $\partial_e K$.

5.5. TRANSFINITE DIAMETER

Evidently $\delta_2(K)$ is just the usual diameter of K, and $\delta_n(K) \leq \delta_2(K)$ for all $n \geq 2$. Indeed, as we shall shortly see, the sequence $(\delta_n(K))_{n \geq 2}$ is decreasing, so it has a limit, often called the *transfinite diameter* of K. Actually, as the following theorem shows, this is nothing other than the capacity.

Theorem 5.5.2 (Fekete–Szegö Theorem) *Let K be a compact subset of $\mathbb{C}$. Then the sequence $(\delta_n(K))_{n \geq 2}$ is decreasing, and*

$$\lim_{n \to \infty} \delta_n(K) = c(K).$$

Proof. In order to simplify the notation, throughout this proof we shall write δ_n for $\delta_n(K)$.

We begin by showing that the sequence $(\delta_n)_{n \geq 2}$ is decreasing. Let $n \geq 2$, and choose $w_1, \ldots, w_{n+1} \in K$ such that

$$\delta_{n+1}^{n(n+1)/2} = \prod_{1 \leq j < k \leq n+1} |w_j - w_k|.$$

Then, since $w_2, \ldots, w_{n+1}$ is an n-tuple in K,

$$\delta_n^{n(n-1)/2} \geq \prod_{2 \leq j < k \leq n+1} |w_j - w_k|.$$

There are $n+1$ such inequalities in all, the m-th one obtained by omitting the terms involving w_m. Multiplying them all together gives

$$\left(\delta_n^{n(n-1)/2}\right)^{n+1} \geq \prod_{1 \leq j < k \leq n+1} |w_j - w_k|^{n-1} = \left(\delta_{n+1}^{n(n+1)/2}\right)^{n-1}.$$

Hence $\delta_n \geq \delta_{n+1}$, as claimed.

Next, we show that $\delta_n \geq c(K)$ for all $n \geq 2$. If $z_1, \ldots, z_n \in K$, then

$$\frac{2}{n(n-1)} \sum_{1 \leq j < k \leq n} \log|z_j - z_k| \leq \log \delta_n.$$

Integrating this inequality with respect to $d\nu(z_1) \cdots d\nu(z_n)$, where ν is an equilibrium measure for K, we obtain

$$\frac{2}{n(n-1)} \sum_{1 \leq j < k \leq n} \int_K \int_K \log|z_j - z_k| \, d\nu(z_j) \, d\nu(z_k) \leq \log \delta_n.$$

Hence $I(\nu) \leq \log \delta_n$, which gives $c(K) \leq \delta_n$, as claimed.

Finally, we show that $\limsup_{n\to\infty} \delta_n \le c(K)$. Let $\epsilon > 0$, and set
$$K^\epsilon = \{z \in \mathbb{C} : \operatorname{dist}(z, K) \le \epsilon\}.$$
Let $n \ge 2$, and choose $w_1, \ldots, w_n \in K$ such that
$$\delta_n^{n(n-1)/2} = \prod_{j<k} |w_j - w_k|.$$
For each j, let μ_j be normalized Lebesgue measure on the circle $\partial \Delta(w_j, \epsilon)$, and put $\mu = n^{-1} \sum_1^n \mu_j$. Then $I(\mu)$ is given by
$$\frac{1}{n^2} \sum_j \iint \log|z-w|\, d\mu_j(z)\, d\mu_j(w) + \frac{2}{n^2} \sum_{j<k} \iint \log|z-w|\, d\mu_j(z)\, d\mu_k(w).$$
Now for each j,
$$\iint \log|z-w|\, d\mu_j(z)\, d\mu_j(w) = I(\mu_j) = \log \epsilon,$$
by Theorem 3.7.7 and Corollary 5.2.2. Also, for each pair $j < k$,
$$\iint \log|z-w|\, d\mu_j(z)\, d\mu_k(w) = \int p_{\mu_j}(w)\, d\mu_k(w) \ge p_{\mu_j}(w_k),$$
because p_{μ_j} is subharmonic, and
$$p_{\mu_j}(w_k) = \int \log|z - w_k|\, d\mu_j(z) \ge \log|w_j - w_k|,$$
because $\log|z - w_k|$ is subharmonic. Therefore
$$I(\mu) \ge \frac{1}{n^2} \sum_j \log \epsilon + \frac{2}{n^2} \sum_{j<k} \log|w_j - w_k| = \frac{1}{n} \log \epsilon + \frac{n-1}{n} \log \delta_n.$$
Since μ is supported on K^ϵ, it follows that
$$c(K^\epsilon) \ge \epsilon^{1/n} \delta_n(K)^{(n-1)/n}.$$
Hence $\limsup_{n\to\infty} \delta_n \le c(K^\epsilon)$, and as ϵ is arbitrary, the desired conclusion now follows from Theorem 5.1.3 (a). $\square$

Much of the importance of this theorem derives from its connection with polynomial approximation. For several reasons, it is of interest to find monic polynomials $q(z)$ for which the sup-norm on K,
$$\|q\|_K := \sup\{|q(z)| : z \in K\},$$
is relatively small. We now consider one such class.

5.5. TRANSFINITE DIAMETER

Definition 5.5.3 Let K be a compact subset of $\mathbb{C}$, and let $n \geq 2$. A *Fekete polynomial* for K of degree n is a polynomial of the form

$$q(z) = \prod_{1}^{n}(z - w_j),$$

where $w_1, \ldots, w_n$ is a Fekete n-tuple for K.

Theorem 5.5.4 *Let K be a compact subset of $\mathbb{C}$.*

(a) *If q is a monic polynomial of degree $n \geq 1$, then $\|q\|_K^{1/n} \geq c(K)$.*

(b) *If q is a Fekete polynomial of degree $n \geq 2$, then $\|q\|_K^{1/n} \leq \delta_n(K)$.*

Proof. (a) Since $K \subset q^{-1}(\overline{\Delta}(0, \|q\|_K))$, it follows from Theorem 5.2.5 that

$$c(K) \leq c(\overline{\Delta}(0, \|q\|_K))^{1/n} = \|q\|_K^{1/n}.$$

(b) Suppose that $q(z) = \prod_1^n (z - w_i)$, where $w_1, \ldots, w_n$ is a Fekete n-tuple for K. If $z \in K$, then $z, w_1, \ldots, w_n$ is an $(n+1)$-tuple in K, so

$$\prod_{i=1}^{n}|z - w_i| \prod_{j<k}|w_j - w_k| \leq \delta_{n+1}(K)^{n(n+1)/2},$$

and hence

$$|q(z)| \leq \frac{\delta_{n+1}(K)^{n(n+1)/2}}{\delta_n(K)^{n(n-1)/2}} \leq \frac{\delta_n(K)^{n(n+1)/2}}{\delta_n(K)^{n(n-1)/2}} = \delta_n(K)^n.$$

Since z is an arbitrary point of K, this proves the result. □

Combining the last two theorems leads immediately to another characterization of capacity, which makes an interesting contrast with the result of Exercise 5.3.5.

Corollary 5.5.5 *Let K be a compact subset of $\mathbb{C}$, and for each $n \geq 1$ let*

$$m_n(K) = \inf\{\|q\|_K : q \text{ is a monic polynomial of degree } n\}.$$

Then

$$\lim_{n \to \infty} m_n(K)^{1/n} = \inf_{n \geq 1} m_n(K)^{1/n} = c(K). \quad \Box$$

A monic polynomial q of degree n for which $\|q\|_K = m_n(K)$ is called a *Chebyshev polynomial*. It can be shown that it exists and, provided K has at least n points, is unique (see Exercise 4). However, the Fekete polynomials have the advantage that, unlike the Chebyshev polynomials, their zeros always belong to K. As an illustration of this, we now use them to prove a strong form of Lemma 3.5.3.

Theorem 5.5.6 (Evans' Theorem) *Let E be a compact polar set. Then there exists a Borel probability measure μ on E such that $p_\mu(z) = -\infty$ for all $z \in E$.*

Proof. Given $n \geq 2$, let $w_1, \ldots, w_n$ be a Fekete n-tuple for K, and let q_n be the corresponding Fekete polynomial. If μ_n denotes the probability measure on K consisting of $1/n$-masses at $w_1, \ldots, w_n$, then by Theorem 5.5.4 (b)

$$p_{\mu_n}(z) = \sum_1^n \log|z - w_j| = \frac{1}{n}\log|q_n(z)| \leq \log \delta_n(E) \quad (z \in E).$$

Now by Theorem 5.5.2 $\lim_{n\to\infty} \delta_n(E) = c(E) = 0$, so, replacing (μ_n) by a subsequence, we may suppose that $p_{\mu_n} \leq -2^n$ on E for each n. If we set $\mu = \sum_1^\infty 2^{-n}\mu_n$, then μ is a Borel probability measure on K, and

$$p_\mu(z) = \sum_1^\infty 2^{-n} p_{\mu_n}(z) \leq \sum_1^\infty 2^{-n}(-2^n) = -\infty \quad (z \in E).$$

Thus μ has the desired properties. □

Knowledge of $\|q\|_K$ also gives us information about how q behaves off K. If D is a bounded component of $\mathbb{C}_\infty \setminus K$ then $|q(z)| \leq \|q\|_K$ for all $z \in D$ by the maximum principle. The next result tells us what happens when D is the unbounded component. The first part contains the basic inequality, and the second part gives some indication of its sharpness.

Theorem 5.5.7 (Bernstein's Lemma) *Let K be a non-polar compact subset of $\mathbb{C}$, and let D be the component of $\mathbb{C}_\infty \setminus K$ containing ∞.*

(a) *If q is a polynomial of degree $n \geq 1$, then*

$$\left(\frac{|q(z)|}{\|q\|_K}\right)^{1/n} \leq e^{g_D(z,\infty)} \quad (z \in D \setminus \{\infty\}),$$

where g_D denotes the Green's function of D.

(b) *If q is a Fekete polynomial for K of degree $n \geq 2$, then*

$$\left(\frac{|q(z)|}{\|q\|_K}\right)^{1/n} \geq e^{g_D(z,\infty)} \left(\frac{c(K)}{\delta_n(K)}\right)^{\tau_D(z,\infty)} \quad (z \in D \setminus \{\infty\}),$$

where τ_D denotes the Harnack distance for D.

5.5. TRANSFINITE DIAMETER

Proof. (a) Multiplying q by a constant, we can suppose that it is monic. If we define

$$u(z) = \frac{1}{n}\log|q(z)| - \frac{1}{n}\log\|q\|_K - g_D(z,\infty) \quad (z \in D \setminus \{\infty\}),$$

then u is subharmonic on $D \setminus \{\infty\}$. Also

$$u(z) = \log|z| - \frac{1}{n}\log\|q\|_K - \log|z| + \log c(K) + o(1) \quad \text{as } z \to \infty,$$

and therefore setting

$$u(\infty) = \log c(K) - \frac{1}{n}\log\|q\|_K$$

makes u subharmonic on D. Now since $\partial D \subset K$, we have

$$\limsup_{z \to \zeta} u(z) \le \frac{1}{n}\log|q(\zeta)| - \frac{1}{n}\log\|q\|_K \le 0 \quad (\zeta \in \partial D),$$

so by the maximum principle $u \le 0$ on D. This implies the result.

(b) If q is a Fekete polynomial, then in particular all its zeros lie in K, and hence u is actually harmonic on D. Also, from part (a), $u \le 0$ on D, so we may apply Harnack's inequality (see Definition 1.3.4) to $-u$ to obtain

$$u(z) \ge \tau_D(z,\infty)u(\infty) \quad (z \in D).$$

Now by Theorem 5.5.4 (b),

$$u(\infty) = \log c(K) - \frac{1}{n}\log\|q\|_K \ge \log c(K) - \log \delta_n(K).$$

Combining these last two inequalities yields the desired conclusion. □

We end the section with an application to polynomial convexity. A compact subset K of $\mathbb{C}$ is *polynomially convex* if, for each $z \in \mathbb{C} \setminus K$, there exists a polynomial q such that $|q(z)| > \|q\|_K$. This will not be the case if z belongs to a bounded component of $\mathbb{C} \setminus K$, so for K to be polynomially convex it is necessary that $\mathbb{C} \setminus K$ be connected. This condition also turns out to be sufficient. A simple compactness argument then shows that, given an open neighbourhood U of K, there is a finite set of polynomials $q_1, \ldots, q_n$ such that

$$\max_{1 \le j \le n} \frac{|q_j(z)|}{\|q_j\|_K} > 1 \quad (z \in \mathbb{C} \setminus U).$$

What is much less obvious is that in fact one polynomial will do.

Theorem 5.5.8 (Hilbert Lemniscate Theorem) *Let K be a compact subset of $\mathbb{C}$ such that $\mathbb{C}\setminus K$ is connected, and let U be an open neighbourhood of K. Then there exists a polynomial q such that*

$$\frac{|q(z)|}{\|q\|_K} > 1 \quad (z \in \mathbb{C}\setminus U).$$

Proof. We can suppose that K is non-polar, otherwise just adjoin a small line segment in U. Let $D = \mathbb{C}_\infty \setminus K$, and put

$$L = \inf_{z \in \mathbb{C}_\infty \setminus U} g_D(z, \infty) \quad \text{and} \quad M = \sup_{z \in \mathbb{C}_\infty \setminus U} \tau_D(z, \infty),$$

so that $L > 0$ and $M < \infty$. Then by Theorem 5.5.7 (b), if q is a Fekete polynomial for K of degree n,

$$\left(\frac{|q(z)|}{\|q\|_K}\right)^{1/n} \geq e^L \left(\frac{c(K)}{\delta_n(K)}\right)^M \quad (z \in \mathbb{C}\setminus U).$$

Since $\delta_n(K) \to c(K)$ as $n \to \infty$, the right-hand side will exceed 1 for all sufficiently large n. $\square$

Exercises 5.5

1. Let K be a compact subset of $\mathbb{C}$, and let $n \geq 2$. Show that if $w_1, \ldots, w_n$ is a Fekete n-tuple for K, then

$$\delta_n(K)^{n(n-1)/2} = \begin{vmatrix} 1 & 1 & \cdots & 1 \\ w_1 & w_2 & \cdots & w_n \\ w_1^2 & w_2^2 & \cdots & w_n^2 \\ \vdots & \vdots & & \vdots \\ w_1^{n-1} & w_2^{n-1} & \cdots & w_n^{n-1} \end{vmatrix}.$$

   Use Hadamard's inequality to deduce that, if K is the unit disc, then

$$\delta_n(K) = n^{1/(n-1)}.$$

2. Use Theorem 5.5.2 to give another proof that the Cantor set has capacity at least $1/9$.

3. Let K be a compact subset of $\mathbb{C}$, let $n \geq 2$, and let $w_1, \ldots, w_n$ be a Fekete n-tuple for K. Show that if q is a polynomial of degree at most $n - 1$, then

$$\|q\|_K \leq \sum_1^n |q(w_j)|.$$

   [Hint: consider first the case when $q(w_1) = 1$ and $q(w_j) = 0$ $(j > 1)$.]

5.5. TRANSFINITE DIAMETER

4. Let K be a compact subset of $\mathbb{C}$ containing at least $n+1$ points, and define
$$m_n(K) = \inf\{\|q\|_K : q \text{ is a monic polynomial of degree } n\}.$$
   (i) Show that the infimum is attained by some polynomial t_n.
   (ii) Show that $|t_n(z)| = m_n(K)$ for at least $n+1$ points of K. [Hint: If not, then there is a polynomial q of degree at most $n-1$ which equals $t_n(z)$ at each $z \in K$ where $|t_n(z)| = m_n(K)$. Show that if $\epsilon > 0$ is chosen sufficiently small, then $\|t_n - \epsilon q\|_K < m_n(K)$.]
   (iii) Deduce that t_n is unique. [Hint: If there are two candidates for t_n, apply (ii) to their average.]
   (iv) Show that all the the zeros of t_n lie in the convex hull of K.
   (v) Suppose that $K = [-1, 1] \times \{i, -i\}$. Show that if n is odd then t_n has a zero on the real axis (and thus outside K).

5. Let K be a non-polar compact subset of $\mathbb{C}$. For each $n \geq 2$, let μ_n be a probability measure on K consisting of $1/n$-masses at the points of a Fekete n-tuple for K. Prove that $\mu_n \xrightarrow{w^*} \nu$ as $n \to \infty$, where ν is the equilibrium measure for K.
   [Hint: Show that if μ is a weak*-limit point of (μ_n), then we have $p_\mu(z) = p_\nu(z)$ for all z in the unbounded component of $\mathbb{C} \setminus K$, and hence for all $z \in \partial_e K$. Deduce that $I(\mu) = I(\nu)$.]

6. Let $\gamma \not\equiv 0$ be a set-function defined on the non-empty compact subsets of $\mathbb{C}$, satisfying the following conditions:
   (a) if $K \subset L$ then $\gamma(K) \leq \gamma(L)$;
   (b) if $K_n \downarrow K$ then $\gamma(K_n) \downarrow \gamma(K)$;
   (c) if $q(z) = \sum_0^n a_j z^j$ then $\gamma(q^{-1}(K)) = (\gamma(K)/|a_n|)^{1/n}$.

   The aim of this exercise is to show that γ is exactly the capacity.
   (i) Show that $\gamma(\overline{\Delta}(0, 1)) = 1$.
   (ii) Show that if $\mathbb{C} \setminus K$ is connected, then there exist $K_n \downarrow K$ such that each K_n has the form $q^{-1}(\overline{\Delta}(0, 1))$ for some polynomial q, and deduce that $\gamma(K) = c(K)$.
   (iii) Show that if $\mathbb{C} \setminus K$ has finitely many components, then there exist $K_n \uparrow K$ such that each $\mathbb{C} \setminus K_n$ is connected, and deduce that $\gamma(K) = c(K)$.
   (iv) Show that if K is an arbitrary compact subset of $\mathbb{C}$, then there exist $K_n \downarrow K$ such that each $\mathbb{C} \setminus K_n$ has finitely many components, and deduce that $\gamma(K) = c(K)$.

Notes on Chapter 5

§5.1

Given a set $E \subset \mathbb{C}$, its *outer capacity* is defined by

$$c^*(E) = \inf\{c(U): \text{ open } U \supset E\},$$

and E is said to be *capacitable* if $c^*(E) = c(E)$. Choquet's theorem (equation (5.3)) amounts to saying that bounded Borel sets are capacitable. What makes this hard to prove is that capacitable sets do not form a σ-algebra—they are not closed under complements—and instead one has to proceed in a less direct manner, via the notion of analytic sets. For more details see [34, §5.8].

Many of the results in this chapter, though proved only for F_σ or Borel sets, remain true for arbitrary sets provided one replaces capacity c by outer capacity c^*. Also, it is common to define polar sets as sets of outer capacity zero, rather than simply capacity zero as we have done. This has the advantage of avoiding pathologies such as Exercise 3.2.2, though it is more complicated to use in practice. In any case, for Borel sets the two definitions coincide, thanks to Choquet's theorem.

Exercise 3 was suggested to the author by Brian Cole.

§5.2

Table 5.1 was compiled from various sources, including [35, 40, 48].

§5.3

Theorem 5.3.2(d) is taken from [51, Chapter 9], which also contains much more about estimating the capacity of subsets of the circle, particularly via the notion of extremal length.

The Ahlfors–Beurling inequality (Lemma 5.3.6), which is from [1], has many other applications. A particularly nice one appears in [2].

Exercise 3(i) is a result of Hayman [33].

Exercise 5 is based on results of Pólya and Szegö [48]. The first inequality in (iv) remains true even if K is disconnected (Theorem 5.3.5), but whether the second inequality still holds appears to be an open problem.

The isoperimetric inequality in Exercise 6 can be improved to

$$\sum_{k \in \mathbb{Z}} k^2 \, \text{area}(U_k) \leq \frac{\text{length}(\gamma)^2}{4\pi}.$$

A proof of this can be found in [10].

Chapter 6

Applications

We have already seen, scattered throughout the book, a multitude of applications of potential theory to complex analysis. In this final chapter we take a look at some applications to other areas: functional analysis, harmonic analysis, approximation theory and dynamical systems. Inevitably, this means that some background has to be taken for granted. A summary of the basic facts needed is provided, together with references where one can pursue the details. The main point is to show how the subharmonic techniques that we have developed can be applied to obtain a wide variety of results, some of them classical, others relatively new. The sections can be read independently of one another.

6.1 Interpolation of L^p-spaces

Let (Ω, μ) be a measure space. Given a measurable function $f\colon \Omega \to \mathbb{C}$ and a number $p \in [1, \infty]$, we define

$$\|f\|_p = \begin{cases} \left(\int_\Omega |f|^p \, d\mu\right)^{1/p} & \text{if } 1 \le p < \infty \\ \operatorname{ess\,sup} |f| & \text{if } p = \infty. \end{cases}$$

By definition, $L^p(\mu)$ is the set of all measurable functions $f\colon \Omega \to \mathbb{C}$ such that $\|f\|_p < \infty$. Identifying functions which agree almost everywhere, the pair $(L^p(\mu), \|\cdot\|_p)$ becomes a Banach space.

There are several relations between the various L^p-norms, the most important of which is Hölder's inequality: if p, p' are conjugate indices (i.e. $1/p + 1/p' = 1$), and if $f \in L^p(\mu)$ and $g \in L^{p'}(\mu)$, then $fg \in L^1(\mu)$, and

$$\|fg\|_1 \le \|f\|_p \|g\|_{p'}.$$

Thus, given $f \in L^p(\mu)$, the linear functional $g \mapsto \int_\Omega fg\, d\mu$ is continuous on $L^{p'}(\mu)$ with norm at most $\|f\|_p$. Moreover, if $1 < p < \infty$, then it has norm exactly $\|f\|_p$, and every continuous linear functional on $L^{p'}(\mu)$ is of this form.

For further information about L^p-spaces see for example [57].

It often happens that one has a linear map T, defined on a space of measurable functions, which is bounded as an operator both on (say) L^1 and on L^2. By Hölder's inequality the sum $L^1 + L^2$ contains all the L^p-spaces for $1 \leq p \leq 2$, and so it makes sense to ask whether T is also bounded as an operator on L^p for each such p. The pleasing answer is yes. It is a special case of the following interpolation theorem which, at first sight, appears to have little to do with complex analysis or potential theory.

Theorem 6.1.1 (Riesz–Thorin Interpolation Theorem) *Let (Ω, μ) and (Σ, ν) be measure spaces, and let T be a linear map from $L^{p_0}(\mu) + L^{p_1}(\mu)$ to $L^{q_0}(\nu) + L^{q_1}(\nu)$, where $p_0, p_1, q_0, q_1 \in [1, \infty]$. If*

$$T\colon L^{p_0}(\mu) \to L^{q_0}(\nu) \quad \text{with norm } \leq M_0$$
$$T\colon L^{p_1}(\mu) \to L^{q_1}(\nu) \quad \text{with norm } \leq M_1,$$

then for each $\theta \in (0,1)$,

$$T\colon L^{p_\theta}(\mu) \to L^{q_\theta}(\nu) \quad \text{with norm } \leq M_0^{1-\theta} M_1^\theta,$$

where p_θ and q_θ are given by

$$\frac{1}{p_\theta} = \frac{1-\theta}{p_0} + \frac{\theta}{p_1} \quad \text{and} \quad \frac{1}{q_\theta} = \frac{1-\theta}{q_0} + \frac{\theta}{q_1}.$$

Proof. Fix $\theta \in (0,1)$. We shall carry out the proof under the assumption that $p_\theta, q_\theta \in (1, \infty)$, indicating at the end the modifications needed to handle the remaining cases. This has a twofold advantage: firstly, $L^{q_\theta}(\nu)$ can then be identified with the dual space of $L^{q'_\theta}(\nu)$ (where, as above, q' denotes the index conjugate to q), and secondly, the simple functions are dense in $L^{p_\theta}(\mu)$ and $L^{q_\theta}(\nu)$. Here, by simple function is meant one of the form $\sum_{j=1}^k c_j 1_{A_j}$, where the c_j are complex numbers and the A_j are disjoint measurable sets of finite measure. Note that a simple function belongs to L^p for every p.

Let ϕ be a simple function on (Ω, μ). From the remarks above about duality and density, it follows that

(6.1) $$\|T\phi\|_{q_\theta} = \sup\left|\int_\Sigma (T\phi)\psi\, d\nu\right|,$$

6.1. INTERPOLATION OF L^p-SPACES

the supremum being taken over all simple functions ψ on (Σ, ν) such that $\|\psi\|_{q_\theta'} \le 1$. Fix such a ψ. Let S be the strip $\{z \in \mathbb{C} : 0 < \operatorname{Re} z < 1\}$, and for $z \in \overline{S}$, define simple functions ϕ_z and ψ_z by

$$\phi_z = \frac{\phi}{|\phi|}|\phi|^{p_\theta\left(\frac{1-z}{p_0}+\frac{z}{p_1}\right)}, \qquad \psi_z = \frac{\psi}{|\psi|}|\psi|^{q_\theta'\left(\frac{1-z}{q_0'}+\frac{z}{q_1'}\right)},$$

and set

$$F(z) = \int_\Sigma (T\phi_z)\psi_z \, d\nu.$$

Then F is a bounded continuous function on $\overline{S}$ which is holomorphic on S. Moreover, if $\operatorname{Re} \zeta = 0$, then by Hölder's inequality

$$|F(\zeta)| \le \|T\phi_\zeta\|_{q_0}\|\psi_\zeta\|_{q_0'} \le M_0\|\phi_\zeta\|_{p_0}\|\psi_\zeta\|_{q_0'} \le M_0\|\phi\|_{p_\theta}^{p_\theta/p_0},$$

while if $\operatorname{Re} \zeta = 1$, then

$$|F(\zeta)| \le \|T\phi_\zeta\|_{q_1}\|\psi_\zeta\|_{q_1'} \le M_1\|\phi_\zeta\|_{p_1}\|\psi_\zeta\|_{q_1'} \le M_1\|\phi\|_{p_\theta}^{p_\theta/p_1}.$$

Hence, by the three-lines theorem (Theorem 2.3.6) applied to $u := \log|F|$, we deduce that

$$|F(\theta)| \le \left(M_0\|\phi\|_{p_\theta}^{p_\theta/p_0}\right)^{1-\theta}\left(M_1\|\phi\|_{p_\theta}^{p_\theta/p_1}\right)^\theta = M_0^{1-\theta}M_1^\theta\|\phi\|_{p_\theta}.$$

From the definition of F, this gives

$$\left|\int_\Sigma (T\phi)\psi \, d\nu\right| \le M_0^{1-\theta}M_1^\theta\|\phi\|_{p_\theta}.$$

As this holds for each such ψ, it follows from (6.1) that

(6.2) $$\|T\phi\|_{q_\theta} \le M_0^{1-\theta}M_1^\theta\|\phi\|_{p_\theta}.$$

To complete the proof, we need to show that the inequality (6.2) holds, not only for simple functions ϕ, but for every $f \in L^{p_\theta}(\mu)$. Given such an f, choose simple functions (ϕ_n) on (Ω, μ) such that $\|\phi_n - f\|_{p_\theta} \to 0$. From (6.2) it follows that $(T\phi_n)$ is a sequence which is Cauchy and hence convergent in $L^{q_\theta}(\nu)$ to some function g satisfying

$$\|g\|_{q_\theta} \le M_0^{1-\theta}M_1^\theta\|f\|_{p_\theta}.$$

It remains to show that $Tf = g$. This is clear if $p_\theta = p_0$ or p_1, since T is then continuous on $L^{p_\theta}(\mu)$, but in general we do not know this *a priori* (indeed, it is the point of the theorem), and so we proceed as follows. We can suppose that $p_0 < p_\theta < p_1$. Let (ϵ_n) be a sequence of positive numbers, to be chosen shortly, and set

$$A_n = \{\omega \in \Omega : |\phi_n(\omega) - f(\omega)| > \epsilon_n\}.$$

By Chebyshev's inequality,

$$\mu(A_n) \leq \left(\|\phi_n - f\|_{p_\theta}/\epsilon_n\right)^{p_\theta},$$

and hence by Hölder's inequality,

(6.3) $$\|(\phi_n - f)1_{A_n}\|_{p_0}^{p_0} \leq \|\phi_n - f\|_{p_\theta}^{p_\theta}/\epsilon_n^{p_\theta - p_0}.$$

On the other hand, on $\Omega \setminus A_n$ we have $\epsilon_n^{-1}|\phi_n - f| \leq 1$, so

$$\|\epsilon_n^{-1}(\phi_n - f)1_{\Omega \setminus A_n}\|_{p_1}^{p_1} \leq \|\epsilon_n^{-1}(\phi_n - f)1_{\Omega \setminus A_n}\|_{p_\theta}^{p_\theta},$$

and hence

(6.4) $$\|(\phi_n - f)1_{\Omega \setminus A_n}\|_{p_1} \leq \epsilon_n^{1-p_\theta/p_1}\|\phi_n - f\|_{p_\theta}^{p_\theta/p_1}.$$

(This reasoning assumes that $p_1 < \infty$, but the final inequality is clearly also true if $p_1 = \infty$.) The right-hand side of (6.4) tends to zero if $\epsilon_n \to 0$, and so does that of (6.3) provided that $\epsilon_n \to 0$ sufficiently slowly. With the sequence (ϵ_n) so chosen, since T is continuous on $L^{p_0}(\mu)$ and $L^{p_1}(\mu)$, it follows that $T((\phi_n - f)1_{A_n}) \to 0$ in $L^{q_0}(\nu)$ and $T((\phi_n - f)1_{\Omega \setminus A_n}) \to 0$ in $L^{q_1}(\nu)$. In particular, both sequences tend to zero in measure and so, adding them together, we deduce that $T\phi_n \to Tf$ in measure. Since we already know that $T\phi_n \to g$ in $L^{q_\theta}(\nu)$, we conclude that $Tf = g$, as desired.

It remains only to take care of the exceptional cases excluded at the beginning of the proof. Suppose first that $1 < p_\theta < \infty$, but now $q_\theta = 1$ or ∞, so that necessarily $q_\theta = q_0 = q_1$. Then there is no need to vary q, and so we can repeat the proof above with F given by

$$F(z) = \Psi(T\phi_z),$$

where Ψ is a fixed linear functional on $L^{q_\theta}(\nu)$ of norm 1. The rôle of (6.1) is now played by the Hahn–Banach theorem.

The other case, where $p_\theta = 1$ or ∞, cannot be treated by making such a simple modification, but here, fortunately, the whole theorem is a direct consequence of Hölder's inequality. Indeed, given $f \in L^{p_\theta}(\mu)$, we have

$$\begin{aligned}
\|Tf\|_{q_\theta} &= \| |Tf|^{1-\theta}|Tf|^\theta \|_{q_\theta} \\
&\leq \|Tf\|_{q_0}^{1-\theta}\|Tf\|_{q_1}^\theta \\
&\leq (M_0^{1-\theta}\|f\|_{p_0}^{1-\theta})(M_1^\theta\|f\|_{p_1}^\theta) \\
&= M_0^{1-\theta}M_1^\theta\|f\|_{p_\theta},
\end{aligned}$$

the final step following from the fact that necessarily $p_\theta = p_0 = p_1$. $\square$

6.1. INTERPOLATION OF L^p-SPACES

Here are two simple applications to illustrate the theorem. The first one is to Fourier series. Let $\mathbb{T}$ denote the unit circle, thought of as a measure space with respect to normalized Lebesgue measure $d\theta/2\pi$. Given $f \in L^1(\mathbb{T})$, its Fourier coefficients are defined by

$$\widehat{f}(n) = \frac{1}{2\pi} \int_0^{2\pi} f(e^{i\theta}) e^{-in\theta}\, d\theta \quad (n \in \mathbb{Z}).$$

Corollary 6.1.2 (Hausdorff–Young Theorem) *If $f \in L^p(\mathbb{T})$, where $1 \leq p \leq 2$, then the sequence $(\widehat{f}(n))$ belongs to $\ell^{p'}(\mathbb{Z})$, where p' is the index conjugate to p, and*

$$\|(\widehat{f}(n))\|_{p'} \leq \|f\|_p.$$

Proof. Certainly, if $f \in L^1(\mathbb{T})$ then

$$\|(\widehat{f}(n))\|_\infty \leq \|f\|_1,$$

and also, by Bessel's inequality, if $f \in L^2(\mathbb{T})$ then

$$\|(\widehat{f}(n))\|_2 \leq \|f\|_2.$$

The result follows by applying the Riesz–Thorin theorem to the linear map $T: f \mapsto (\widehat{f}(n))$. $\square$

The second application is to convolutions. If $f, g: \mathbb{R} \to \mathbb{C}$ are measurable functions, then their convolution is defined by

$$f * g(x) = \int_{-\infty}^\infty f(x-y) g(y)\, dy,$$

whenever this integral exists.

Corollary 6.1.3 (Young's Inequality) *If $f \in L^p(\mathbb{R})$ and $g \in L^q(\mathbb{R})$, where $1/p + 1/q \geq 1$, then $f * g \in L^r(\mathbb{R})$, where $1/r = 1/p + 1/q - 1$, and*

$$\|f * g\|_r \leq \|f\|_p \|g\|_q.$$

Proof. Fix $p \in [1, \infty]$ and $f \in L^p(\mathbb{R})$. If $g \in L^{p'}(\mathbb{R})$, then by Hölder's inequality

$$|f * g(x)| \leq \left(\int_{-\infty}^\infty |f(x-y)|^p\, dy \right)^{1/p} \left(\int_{-\infty}^\infty |g(y)|^{p'}\, dy \right)^{1/p'},$$

so $f * g \in L^\infty(\mathbb{R})$ and

$$\|f * g\|_\infty \leq \|f\|_p \|g\|_{p'}.$$

On the other hand, if $g \in L^1(\mathbb{R})$, then Hölder's inequality gives

$$|f * g(x)| \leq \left(\int_{-\infty}^{\infty} |f(x-y)|^p |g(y)|\, dy\right)^{1/p} \left(\int_{-\infty}^{\infty} |g(y)|\, dy\right)^{1/p'},$$

so taking pth powers and integrating with respect to x we obtain

$$\|f * g\|_p \leq \|f\|_p \|g\|_1.$$

The result now follows by applying the Riesz–Thorin theorem to the linear map $T: g \mapsto f * g$. □

6.2 Homogeneous Polynomials

A polynomial $p(z_1, \ldots, z_n)$ on $\mathbb{C}^n$ is said to be *homogeneous of degree d* if

$$p(\lambda z_1, \ldots, \lambda z_n) = \lambda^d p(z_1, \ldots, z_n) \qquad (\lambda, z_1, \ldots, z_n \in \mathbb{C}).$$

In this section we prove a symmetrization inequality for such polynomials which has some interesting links with capacity. As in the previous section, the result hinges on the convexity properties of subharmonic functions, but this time applied in rather a different way.

Theorem 6.2.1 *Let p be a homogeneous polynomial on $\mathbb{C}^n$. If $E_1, \ldots, E_n$ are non-empty subsets of $\mathbb{C}$, then*

(6.5) $$\sup |p|(E_1, \ldots, E_n) \geq \sup |p|(\overline{\Delta}_1, \ldots, \overline{\Delta}_n),$$

where $\overline{\Delta}_j = \overline{\Delta}(0, c(E_j))$.

Here we have used the abbreviation

$$\sup |p|(E_1, \ldots, E_n) := \sup\{|p(z_1, \ldots, z_n)| : z_1 \in E_1, \ldots, z_n \in E_n\}.$$

Theorem 6.2.1 says that this quantity is always decreased by replacing each set E_j by a disc around 0 of the same capacity. In the case $c(E_j) = \infty$, the disc $\overline{\Delta}_j$ is the whole of $\mathbb{C}$, and so we immediately obtain the following corollary.

Corollary 6.2.2 *If p is bounded on $E_1 \times \cdots \times E_n$, where $c(E_j) = \infty$ for each j, then p is bounded on $\mathbb{C}^n$, and hence constant.* □

6.2. HOMOGENEOUS POLYNOMIALS

We shall derive some further consequences after the proof of the theorem. First, we need a lemma.

Lemma 6.2.3 *Let p be a polynomial on $\mathbb{C}^n$, let $q_1, \ldots, q_n$ be polynomials on $\mathbb{C}$ of degrees $d_1, \ldots, d_n \geq 1$ respectively, and define $\phi: (0, \infty) \to \mathbb{R}$ by*

$$\phi(r) = \sup\{\log |p(z_1, \ldots, z_n)| : |q_1(z_1)|^{1/d_1} \leq r, \ldots, |q_n(z_n)|^{1/d_n} \leq r\}.$$

Then $\phi(r)$ is a convex function of $\log r$.

Proof. Define $u: \mathbb{C} \to [-\infty, \infty)$ by

$$u(\lambda) = \sup\{\log |p(z_1, \ldots, z_n)| : q_1(z_1) = \lambda^{d_1}, \ldots, q_n(z_n) = \lambda^{d_n}\}.$$

Then certainly u is continuous on $\mathbb{C}$. Also, if F is the finite set given by

$$F = \bigcup_{j=1}^{n} \{\lambda \in \mathbb{C} : \lambda^{d_j} = q_j(z) \text{ where } q_j'(z) = 0\},$$

then each point of $\mathbb{C} \setminus F$ has an open neighbourhood on which $u(\lambda)$ is the maximum of a finite number of functions of the form

$$\log |p(f_1(\lambda^{d_1}), \ldots, f_n(\lambda^{d_n}))|,$$

each f_j being a local holomorphic inverse of q_j. Therefore u is subharmonic on $\mathbb{C} \setminus F$, and as it is continuous on F it follows from the removable singularity theorem (Theorem 3.0.1) that it is actually subharmonic on the whole of $\mathbb{C}$. Finally, since $\phi(r) = \sup_{|\lambda| \leq r} u(\lambda)$, we deduce from Theorem 2.6.8 (a) that $\phi(r)$ is a convex function of $\log r$. $\square$

Proof of Theorem 6.2.1. We begin by making some reductions. Using Theorem 5.1.2 (b), it suffices to prove the result in the case when the sets $E_1, \ldots, E_n$ are compact. Moreover, we can also suppose that they have connected complements in $\mathbb{C}$, because filling in the holes does not affect either side of (6.5). Finally, by Hilbert's lemniscate theorem (Theorem 5.5.8), each E_j can be approximated arbitrarily closely on the outside by lemniscates, and so we may as well assume that

$$E_j = \{z : |q_j(z)|^{1/d_j} \leq c_j\} \quad (j = 1, \ldots, n),$$

where q_j is a monic polynomial of degree d_j, say, and $c_j > 0$. Note that this implies that $c(E_j) = c_j$, by Theorem 5.2.5.

Now, for $r > 0$ set

$$E_j(r) = \{z : |q_j(z)|^{1/d_j} \leq c_j r\},$$
$$\phi(r) = \sup \log |p|(E_1(r), \ldots, E_n(r)).$$

By Lemma 6.2.3 $\phi(r)$ is a convex function of $\log r$. Also, defining
$$\overline{\Delta}_j(r) = \{z : |z| \le c_j r\},$$
$$\psi(r) = \sup \log |p|(\overline{\Delta}_1(r), \ldots, \overline{\Delta}_n(r)),$$
it follows from the homogeneity of p that $\psi(r) = \psi(1) + d \log r$, where d is the degree of homogeneity. Thus $\phi(r) - \psi(r)$ is still a convex function of $\log r$.

Next, we show that

(6.6) $$\phi(r) - \psi(r) \to 0 \quad \text{as } r \to \infty.$$

To see this, for each j choose R_j large enough so that q_j has all its zeros in $\{|z| \le R_j\}$. Since q_j is monic of degree d_j, an elementary estimate shows that
$$|z| + R_j \ge |q_j(z)|^{1/d_j} \ge |z| - R_j \quad (|z| > R_j),$$
and hence
$$\overline{\Delta}_j(r - R_j/c_j) \subset E_j(r) \subset \overline{\Delta}_j(r + R_j/c_j) \quad (r > R_j/c_j).$$
Thus, if $r_0 = \max_j (R_j/c_j)$, then
$$\psi(r - r_0) \le \phi(r) \le \psi(r + r_0) \quad (r > r_0),$$
from which (6.6) follows, since $\psi(r + r_0) - \psi(r - r_0) \to 0$ as $r \to \infty$.

Finally, a convex function which tends to zero at infinity is necessarily positive. Therefore $\phi - \psi \ge 0$, and in particular $\phi(1) \ge \psi(1)$, which gives the inequality (6.5). $\square$

Here are two applications of this theorem to capacity. The first is a sharpening of the inequality $c(K) \le \delta_n(K)$ between capacity and n-th diameter (see Theorem 5.5.2).

Corollary 6.2.4 *If K is a compact subset of $\mathbb{C}$ and $n \ge 2$, then*
$$c(K) \le \delta_n(K)/n^{1/(n-1)}.$$

Proof. Applying Theorem 6.2.1 with
$$p(z_1, \ldots, z_n) = \prod_{j<k} (z_j - z_k)$$
and $E_1 = \cdots = E_n = K$, we deduce that
$$\delta_n(K) \ge \delta_n\bigl(\overline{\Delta}(0, c(K))\bigr) = c(K) \delta_n\bigl(\overline{\Delta}(0, 1)\bigr).$$

It remains only to evaluate the constant $\delta_n\bigl(\overline{\Delta}(0, 1)\bigr)$, and this was shown to be $n^{1/(n-1)}$ in Exercise 5.5.1. $\square$

6.3. UNIFORM APPROXIMATION

The second application contains, as a special case, the fact that the Minkowski sum and product of compact sets K_1, K_2 satisfy

$$c(K_1 + K_2) \geq c(K_1) + c(K_2) \quad \text{and} \quad c(K_1 K_2) \geq c(K_1) c(K_2).$$

Corollary 6.2.5 *If p is a homogeneous polynomial on $\mathbb{C}^n$, and $K_1, \ldots, K_n$ are compact subsets of $\mathbb{C}$, then*

$$c\bigl(p(K_1, \ldots, K_n)\bigr) \geq \bigl|p\bigl(c(K_1), \ldots, c(K_n)\bigr)\bigr|.$$

Proof. Applying Theorem 6.2.1 to the homogeneous polynomial on $\mathbb{C}^{mn}$ given by

$$\prod_{1 \leq j < k \leq m} \bigl(p(z_{j,1}, \ldots, z_{j,n}) - p(z_{k,1}, \ldots, z_{k,n})\bigr),$$

with $E_{r,s} = K_s$ ($1 \leq r \leq m$, $1 \leq s \leq n$), we obtain

$$\delta_m\bigl(p(K_1, \ldots, K_n)\bigr) \geq \delta_m\bigl(p(\overline{\Delta}_1, \ldots, \overline{\Delta}_n)\bigr),$$

where $\overline{\Delta}_j = \overline{\Delta}(0, c(K_j))$. Letting $m \to \infty$, we get

$$c\bigl(p(K_1, \ldots, K_n)\bigr) \geq c\bigl(p(\overline{\Delta}_1, \ldots, \overline{\Delta}_n)\bigr).$$

Now as p is homogeneous, the set $p(\overline{\Delta}_1, \ldots, \overline{\Delta}_n)$ is a disc whose radius (and thus capacity) is at least $|p(c(K_1), \ldots, c(K_n))|$. In conjunction with the last inequality, this gives the result. □

6.3 Uniform Approximation

Let K be a compact subset of $\mathbb{C}$, and $f : K \to \mathbb{C}$ be a continuous function. A general problem in approximation theory is to determine whether, given a class $\mathcal{C}$ of continuous functions on K, it is possible to find a sequence (f_n) in $\mathcal{C}$ such that $f_n \to f$ uniformly on K. The Stone–Weierstrass theorem tells us that the answer is always yes if $\mathcal{C}$ is an algebra which is self-adjoint and separates points. However, there are many interesting cases where either $\mathcal{C}$ is not self-adjoint, or it fails to be an algebra. We shall look at one of each.

For our first example, consider what happens when the approximating class $\mathcal{C}$ is the algebra of polynomials. Note that $\mathcal{C}$ is not self-adjoint. The classical theorem of Runge asserts that f can be approximated uniformly on K by polynomials provided that $\mathbb{C} \setminus K$ is connected and f extends to be holomorphic on a neighbourhood of K. With the aid of a little potential theory, we can prove a quantitative version of this result which is somewhat stronger. For $n \geq 1$, we write

$$d_n(f, K) = \inf\{\|f - p\|_K : p \text{ is a polynomial of degree at most } n\}.$$

Theorem 6.3.1 (Bernstein–Walsh Theorem) *Let K be a compact subset of $\mathbb{C}$ such that $\mathbb{C} \setminus K$ is connected. If f is a function holomorphic on an open neighbourhood U of K, then*

(6.7) $$\limsup_{n \to \infty} d_n(f, K)^{1/n} \leq \theta < 1,$$

where

(6.8) $$\theta = \begin{cases} \sup_{\mathbb{C}_\infty \setminus U} \exp(-g_{\mathbb{C}_\infty \setminus K}(z, \infty)) & \text{if } c(K) > 0, \\ 0 & \text{if } c(K) = 0. \end{cases}$$

Proof. Suppose first that $c(K) > 0$. Let Γ be a closed contour in $U \setminus K$ which winds once around each point of K and zero times round each point of $\mathbb{C} \setminus U$. Given $n \geq 2$, let q_n be a Fekete polynomial of degree n for K (see Definition 5.5.3), and define

$$p_n(w) = \frac{1}{2\pi i} \int_\Gamma \frac{f(z)}{q_n(z)} \frac{q_n(w) - q_n(z)}{w - z} \, dz.$$

Clearly p_n is a polynomial of degree at most $n - 1$. Also, applying Cauchy's integral formula to f and subtracting gives

(6.9) $$f(w) - p_n(w) = \frac{1}{2\pi i} \int_\Gamma \frac{f(z)}{z - w} \frac{q_n(w)}{q_n(z)} \, dz \quad (w \in K).$$

Consequently,

$$d_n(f, K) \leq \|f - p_n\|_K \leq C \frac{\sup_K |q_n|}{\inf_\Gamma |q_n|},$$

where C is a constant independent of n. Now by Theorem 5.5.7 (b),

$$\left(\frac{\sup_K |q_n|}{\inf_\Gamma |q_n|}\right)^{1/n} \leq \alpha \left(\frac{\delta_n(K)}{c(K)}\right)^\beta,$$

where

$$\alpha = \sup_\Gamma \exp(-g_{\mathbb{C}_\infty \setminus K}(z, \infty)) \quad \text{and} \quad \beta = \sup_\Gamma \tau_{\mathbb{C}_\infty \setminus K}(z, \infty).$$

Hence

$$\limsup_{n \to \infty} d_n(f, K)^{1/n} \leq \limsup_{n \to \infty} C^{1/n} \alpha \left(\frac{\delta_n(K)}{c(K)}\right)^\beta = \alpha.$$

Finally, since Γ can be chosen to lie outside any preassigned compact subset of U, we can make α as close as we please to the number θ defined in (6.8), which gives (6.7).

6.3. UNIFORM APPROXIMATION

Now suppose that $c(K) = 0$. Let $(K_k)_{k \geq 1}$ be a decreasing sequence of non-polar compact subsets of U, with connected complements, such that $K_k \downarrow K$. If θ_k denote the corresponding numbers defined by (6.8), then from what we have already proved,

$$\limsup_{n \to \infty} d_n(f,K)^{1/n} \leq \limsup_{n \to \infty} d_n(f,K_k)^{1/n} \leq \theta_k,$$

and so it suffices to show that $\theta_k \to 0$ as $k \to \infty$. To this end, define

$$h_k(z) = \begin{cases} g_{\mathbb{C}_\infty \setminus K_k}(z, \infty) - g_{\mathbb{C}_\infty \setminus K_1}(z, \infty) & \text{if } z \in \mathbb{C} \setminus K_1, \\ \log c(K_1) - \log c(K_k) & \text{if } z = \infty. \end{cases}$$

Then $(h_k)_{k \geq 1}$ is an increasing sequence of harmonic functions on $\mathbb{C}_\infty \setminus K_1$, and $h_k(\infty) \to \infty$, so by Harnack's theorem $h_k \to \infty$ locally uniformly on $\mathbb{C}_\infty \setminus K_1$. In particular $g_{\mathbb{C}_\infty \setminus K_k}(z, \infty) \to \infty$ uniformly on $\mathbb{C}_\infty \setminus U$, which implies that $\theta_k \to 0$, as desired. □

Notice that if $w_1, \ldots, w_n$ are the zeros of q_n in the proof above (i.e. the Fekete n-tuple for K that gave rise to q_n in the first place), then (9.9) implies that $p_n(w_j) = f(w_j)$ for each j, so that p_n is precisely the polynomial of degree $\leq n-1$ that interpolates f at $w_1, \ldots, w_n$. Even if we had not spotted this, we could still show that $\|f - p_n\|_K \to 0$ by reasoning as follows. If p is any other polynomial of degree $\leq n-1$, then by the result of Exercise 5.5.3,

$$\|p_n - p\|_K \leq \sum_{j=1}^n |p_n(w_j) - p(w_j)| = \sum_{j=1}^n |f(w_j) - p(w_j)| \leq n \|f - p\|_K,$$

whence

$$\|f - p_n\|_K \leq \|f - p\|_K + \|p - p_n\|_K \leq (n+1)\|f - p\|_K.$$

Taking the infimum over all such p, we deduce that

$$\|f - p_n\|_K \leq (n+1) d_n(f, K).$$

By Theorem 6.3.1, the sequence $d_n(f,K)$ tends to 0 sufficiently rapidly to ensure that $\|f - p_n\|_K \to 0$.

The original Runge theorem would not have been strong enough for this purpose. This raises the natural question of whether Theorem 6.3.1 gives the best rate of decay of $d_n(f,K)$. The following converse shows that it does.

Theorem 6.3.2 *Let K be a compact subset of $\mathbb{C}$ such that $\mathbb{C} \setminus K$ is connected, and assume further that K is non-thin at each point of itself. If $f: K \to \mathbb{C}$ is a continuous function such that (6.7) holds for some $\theta < 1$, then f extends holomorphically to an open neighbourhood U of K such that (6.8) holds.*

Proof. The technical condition that K be non-thin at each point of itself guarantees that $c(K) > 0$, and that $g_{\mathbf{C}_\infty \setminus K}(z, \infty)$ tends to zero at each point of ∂K. We can therefore extend $g_{\mathbf{C}_\infty \setminus K}(z, \infty)$ continuously to the whole of $\mathbf{C}_\infty$ by setting it identically equal to zero on K. Let

$$U = \{z \in \mathbf{C}_\infty : \exp(g_{\mathbf{C}_\infty \setminus K}(z, \infty)) < 1/\theta\},$$

so that U is an open neighbourhood of K for which (6.8) holds. By the hypothesis (6.7), there exist polynomials (p_n), with $\deg p_n \leq n$ for each n, such that

$$\limsup_{n \to \infty} \|f - p_n\|_K^{1/n} \leq \theta.$$

We shall show that (p_n) converges locally uniformly on U. If so, then the limit is a holomorphic function on U which agrees with f on K, thereby proving the theorem.

Let L be a compact subset of U. By Theorem 5.5.7 (a),

$$\|p_n - p_{n-1}\|_L \leq \gamma^n \|p_n - p_{n-1}\|_K,$$

where

$$\gamma = \sup_L \exp(g_{\mathbf{C}_\infty \setminus K}(z, \infty)) < 1/\theta.$$

Therefore

$$\limsup_{n \to \infty} \|p_n - p_{n-1}\|_L^{1/n} \leq \limsup_{n \to \infty} \gamma \|p_n - p_{n-1}\|_K^{1/n} \leq \gamma\theta < 1,$$

which implies that $\sum_1^\infty (p_n - p_{n-1})$ converges uniformly on L. In other words, the sequence (p_n) converges uniformly on L, which is what we had to prove. □

Thus, subject to the conditions on K in this theorem, $d_n(f, K) \to 0$ at a geometric rate if and only if f extends holomorphically to a neighbourhood of K. One is led naturally to ask if there is a similar characterization of when $d_n(f, K) \to 0$ (at whatever rate), or in other words, to seek exact conditions for f to be uniformly approximable on K by polynomials. Clearly, for such approximation to be possible, f must be continuous on K and holomorphic on the interior of K. It turns out that, provided as usual that $\mathbf{C} \setminus K$ is connected, this necessary condition on f is also sufficient. This is Mergelyan's theorem: it is altogether more delicate than Runge's theorem and, unfortunately, to prove it here would take us too far afield, though we shall return briefly to the subject at the end of the section.

6.3. UNIFORM APPROXIMATION

We now turn to the other main theorem of this section.

Theorem 6.3.3 (Keldysh's Theorem) *Let K be a compact subset of $\mathbb{C}$ such that $\mathbb{C} \setminus K$ is non-thin at every point of ∂K. Let Λ be a subset of $\mathbb{C} \setminus K$ which contains at least one point from each bounded component of $\mathbb{C} \setminus K$. Then every continuous function $\phi \colon \partial K \to \mathbb{R}$ can be approximated uniformly on ∂K by functions of the form*

$$\text{Re } q(z) + a \log |r(z)|, \tag{6.10}$$

where $a \in \mathbb{R}$, and q and r are rational functions such that the poles of q and the poles and zeros of r all lie in $\Lambda \cup \{\infty\}$.

Before proving this result, it is worth making some remarks. The first is that the class $\mathcal{C}$ of functions of the form (6.10) is a (real) vector space, but not an algebra, so once again the Stone–Weierstrass theorem is not applicable.

Secondly, the logarithmic term in (6.10) really is necessary. To see this, take K to be the annulus $\{z : 1 \leq |z| \leq 2\}$. If $\phi = \text{Re } q$, where q is a rational function with poles outside K, then by Cauchy's theorem

$$\frac{1}{2\pi} \int_0^{2\pi} \left(\phi(2e^{i\theta}) - \phi(e^{i\theta}) \right) d\theta = \text{Re } \frac{1}{2\pi i} \int_{\partial K} \frac{q(z)}{z} \, dz = 0,$$

and hence the same is true whenever ϕ is the limit of such functions on ∂K. On the other hand, this integral does not vanish for every continuous function ϕ on ∂K (consider $|z|$ for example).

Thirdly, the theorem is false if the non-thinness condition is omitted (there are 'Swiss-cheese' counter-examples). However, this condition will be automatically satisfied if $\mathbb{C} \setminus K$ has only finitely many components, since every point of ∂K then belongs to the closure of one of these components, and by Theorem 3.8.3 a connected set (other than a singleton) is non-thin at each point of its closure.

In particular this will be true if $\mathbb{C} \setminus K$ is connected, and in this case we may also take Λ to be the empty set, so that q is a polynomial and r must be constant. Theorem 6.3.3 then reads as follows.

Corollary 6.3.4 (Walsh–Lebesgue Theorem) *Let K be a compact subset of $\mathbb{C}$ such that $\mathbb{C} \setminus K$ is connected. Then every continuous function $\phi \colon \partial K \to \mathbb{R}$ can be approximated uniformly on K by functions of the form $\text{Re } q$, where q is a polynomial.* □

We shall derive some further consequences of Theorem 6.3.3 later, but now we turn to the proof, beginning with a lemma.

Lemma 6.3.5 *Let K be as in Theorem 6.3.3. If $w \in \text{int}(K)$, then there exists a Borel probability measure ω on ∂K such that*

$$\log|z - w| = \int_{\partial K} \log|z - \zeta|\, d\omega(\zeta) \quad (z \in \partial K).$$

Proof. Let $\mathcal{H}$ be the space of all continuous functions $h \colon \partial K \to \mathbb{R}$ which extend to be harmonic on a neighbourhood of K. For $h \in \mathcal{H}$, denote by $\epsilon(h)$ the value of the extension of h at w. The maximum principle ensures that this is well-defined, and that $|\epsilon(h)| \leq \|h\|_{\partial K}$. By the Hahn–Banach theorem, ϵ extends to a linear functional on the whole of $C(\partial K)$, satisfying $|\epsilon(f)| \leq \|f\|_{\partial K}$ for all $f \in C(\partial K)$. Clearly $\epsilon(1) = 1$. Also, if $f \geq 0$, then

$$\|f\|_{\partial K} - \epsilon(f) = \epsilon(\|f\|_{\partial K} 1 - f) \leq \big\|\, \|f\|_{\partial K} 1 - f\big\|_{\partial K} \leq \|f\|_{\partial K},$$

so that $\epsilon(f) \geq 0$. Therefore, by the Riesz representation theorem (Theorem A.3.2) there exists a Borel probability measure ω on ∂K such that

$$\epsilon(f) = \int_{\partial K} f\, d\omega \quad (f \in C(\partial K)).$$

Now if $z \in \mathbb{C} \setminus K$ and $h(\zeta) = \log|z - \zeta|$, then $h \in \mathcal{H}$ and $\epsilon(h) = \log|z - w|$. Hence we deduce that

$$\log|z - w| = \int_{\partial K} \log|z - \zeta|\, d\omega(\zeta) \quad (z \in \mathbb{C} \setminus K).$$

Finally, since both sides of this equation are subharmonic functions of z, and $\mathbb{C} \setminus K$ is non-thin at each point of ∂K, it follows that the same equation holds for all $z \in \partial K$. □

Proof of Theorem 6.3.3. We use duality. By the Hahn–Banach theorem, it is enough to prove that every continuous linear functional on $C(\partial K)$ which vanishes on all functions of the form (6.10) must be identically zero. Now a continuous linear functional on $C(\partial K)$ can be expressed as the difference of two positive linear functionals, each of which, by the Riesz representation theorem, is given by a finite Borel measure on ∂K. Thus what we have to prove is the following: if μ_1 and μ_2 are finite Borel measures on ∂K, such that

(6.11) $$\int_{\partial K} (\operatorname{Re} q)\, d\mu_1 = \int_{\partial K} (\operatorname{Re} q)\, d\mu_2$$

for all rational functions q with poles in $\Lambda \cup \{\infty\}$, and also such that

(6.12) $$\int_{\partial K} \log|r|\, d\mu_1 = \int_{\partial K} \log|r|\, d\mu_2$$

for all rational functions r with poles and zeros in $\Lambda \cup \{\infty\}$, then $\mu_1 = \mu_2$.

6.3. UNIFORM APPROXIMATION

Suppose, then, that μ_1, μ_2 satisfy (6.11) and (6.12). If $|w| > \sup_K |\zeta|$, then

$$\int_{\partial K} \log |w - \zeta| \, d\mu_1(\zeta)$$
$$= \int_{\partial K} \log \left| 1 - \frac{\zeta}{w} \right| d\mu_1(\zeta) + \int_{\partial K} \log |w| \, d\mu_1(\zeta)$$
$$= -\sum_{k=1}^{\infty} \frac{1}{k} \int_{\partial K} \operatorname{Re} \left(\frac{\zeta^k}{w^k} \right) d\mu_1(\zeta) + \log |w| \int_{\partial K} 1 \, d\mu_1(\zeta),$$

and similarly for μ_2, so by (6.11),

$$\int_{\partial K} \log |w - \zeta| \, d\mu_1(\zeta) = \int_{\partial K} \log |w - \zeta| \, d\mu_2(\zeta) \qquad (|w| > \sup_K |\zeta|).$$

Also, if $\lambda \in \Lambda$ and $|w - \lambda| < \operatorname{dist}(\lambda, K)$, then

$$\int_{\partial K} \log |w - \zeta| \, d\mu_1(\zeta)$$
$$= \int_{\partial K} \log \left| 1 - \frac{w - \lambda}{\zeta - \lambda} \right| d\mu_1(\zeta) + \int_{\partial K} \log |\zeta - \lambda| \, d\mu_1(\zeta)$$
$$= -\sum_{k=1}^{\infty} \frac{1}{k} \int_{\partial K} \operatorname{Re} \left(\frac{(w - \lambda)^k}{(\zeta - \lambda)^k} \right) d\mu_1(\zeta) + \int_{\partial K} \log |\zeta - \lambda| \, d\mu_1(\zeta),$$

and similarly for μ_2, so by (6.11) and (6.12),

$$\int_{\partial K} \log |w - \zeta| \, d\mu_1(\zeta) = \int_{\partial K} \log |w - \zeta| \, d\mu_2(\zeta) \qquad (|w - \lambda| < \operatorname{dist}(\lambda, K)).$$

What we have shown is that the potentials of μ_1 and μ_2 satisfy $p_{\mu_1} = p_{\mu_2}$ on a non-empty open subset of each component of $\mathbb{C} \setminus K$. As they are both harmonic on $\mathbb{C} \setminus K$, it follows by the identity principle (Theorem 1.1.7) that $p_{\mu_1} = p_{\mu_2}$ everywhere on $\mathbb{C} \setminus K$. Also, as they are both subharmonic on $\mathbb{C}$, and as $\mathbb{C} \setminus K$ is non-thin at every point of ∂K, it follows that $p_{\mu_1} = p_{\mu_2}$ on ∂K. Finally, if $w \in \operatorname{int}(K)$ then, with ω as in Lemma 6.3.5, we have

$$p_{\mu_1}(w) = \int_{\partial K} p_{\mu_1}(\zeta) \, d\omega(\zeta) = \int_{\partial K} p_{\mu_2}(\zeta) \, d\omega(\zeta) = p_{\mu_2}(w),$$

the two outer equalities holding by virtue of Fubini's theorem, and the middle one because $p_{\mu_1} = p_{\mu_2}$ on ∂K. Thus, we have proved that $p_{\mu_1} = p_{\mu_2}$ on the whole of $\mathbb{C}$, and by Corollary 3.7.5 we may therefore conclude that $\mu_1 = \mu_2$, as desired. $\square$

To finish the section, we look at two applications of Theorem 6.3.3. They are both based on the observation that functions of the form (6.10) are harmonic on $\mathbb{C} \setminus \Lambda$, and consequently satisfy the maximum principle on K. The first is the solution of the analogue, in this context, of the Dirichlet problem. Here, the rôle of boundary regularity is played by the condition that $\mathbb{C} \setminus K$ be non-thin at each point of ∂K.

Corollary 6.3.6 *Let K be as in Theorem 6.3.3. Then, given a continuous function $\phi: \partial K \to \mathbb{R}$, there exists a unique continuous function $h: K \to \mathbb{R}$ such that h is harmonic on $\text{int}(K)$ and $h = \phi$ on ∂K.*

Proof. By Theorem 6.3.3, there exist functions (h_n) harmonic on a neighbourhood of K such that $h_n \to \phi$ uniformly on ∂K. Now by the maximum principle, $\|h_n - h_m\|_K = \|h_n - h_m\|_{\partial K}$ for each m, n. Thus, since (h_n) is uniformly Cauchy on ∂K, it must be on K also. Therefore it converges uniformly on K to a function h satisfying all the conclusions of the corollary. The uniqueness of h follows, as usual, from the maximum principle. □

Our second application is to harmonic approximation on K.

Corollary 6.3.7 *Let K and Λ be as in Theorem 6.3.3. If $h: K \to \mathbb{R}$ is continuous on K and harmonic on $\text{int}(K)$, then h can be approximated uniformly on K by functions harmonic on $\mathbb{C} \setminus \Lambda$.*

Proof. By Theorem 6.3.3, there exist functions (h_n) harmonic on $\mathbb{C} \setminus \Lambda$ such that $h_n \to h$ uniformly on ∂K. By the maximum principle, $\|h_n - h\|_K = \|h_n - h\|_{\partial K}$ for each n, so $h_n \to h$ uniformly on K. □

If $\mathbb{C} \setminus K$ is connected then, repeating the argument above using the Walsh–Lebesgue theorem (Corollary 6.3.4), it follows that every function $h: K \to \mathbb{R}$ which is continuous on K and harmonic on $\text{int}(K)$ can be approximated uniformly on K by real parts of polynomials. This is a real-valued analogue of the theorem of Mergelyan mentioned earlier, and indeed some proofs of that theorem make use of Corollary 6.3.4 as an essential step. One such proof can be found in Stout's book [63], which also contains much more information on uniform approximation in general.

6.4 Banach Algebras

A *Banach algebra* is a complex Banach space $(A, \|\cdot\|)$ on which is defined an associative bilinear multiplication satisfying $\|ab\| \leq \|a\|\|b\|$ for all $a, b \in A$. If A has a multiplicative identity e, then we also assume that $\|e\| = 1$. If A has no such identity, then one can always be adjoined to form a Banach algebra $A^{\#}$ in which A is a closed ideal of codimension one.

6.4. BANACH ALGEBRAS

Important examples of Banach algebras include:

(i) the bounded linear operators on a Banach space, with the operator norm, and multiplication given by composition;

(ii) the continuous functions on a compact Hausdorff space, with the sup-norm, and multiplication defined pointwise;

(iii) the integrable functions on a locally compact group, with the L^1-norm, and multiplication given by convolution.

Given an element a of a Banach algebra A, its *spectrum* is defined by

$$\sigma(a) = \{w \in \mathbb{C} : a - we \text{ is not invertible in } A\}.$$

This definition assumes that A has an identity; if it does not, then $\sigma(a)$ is defined relative to $A^\#$ instead. The spectrum is always a non-empty compact set, so we can define the *spectral radius* of a by

$$\rho(a) = \sup\{|w| : w \in \sigma(a)\}.$$

Although $\sigma(a)$, and hence $\rho(a)$, are defined purely in terms of the algebraic structure of A, they are linked to the norm by the spectral radius formula:

$$(6.13) \qquad \rho(a) = \lim_{n \to \infty} \|a^n\|^{1/n} = \inf_{n \geq 1} \|a^n\|^{1/n}.$$

For a proof of this, and more information on Banach algebras in general, see [17, 58].

In this section we shall prove three theorems relating Banach algebras and subharmonic functions, with applications to semisimple, commutative and radical Banach algebras respectively (these terms will be explained later). All three are based, in various ways, on the spectral radius formula (6.13) and the following simple lemma about vector-valued holomorphic functions.

Lemma 6.4.1 *If $f: U \to X$ is a holomorphic function from an open subset U of $\mathbb{C}$ to a complex Banach space $(X, \|\cdot\|)$, then $\log \|f\|$ is a subharmonic function on U.*

Proof. Certainly $\log \|f\|$ is upper semicontinuous on U, indeed even continuous, so we need only check the submean inequality. Let $\overline{\Delta}(z_0, r)$ be a closed disc in U. By the Hahn–Banach theorem, there exists a continuous linear functional ϕ on X of norm 1 such that $\phi(f(z_0)) = \|f(z_0)\|$. Then $\phi \circ f$

is a scalar-valued holomorphic function on U, so $\log|\phi \circ f|$ is subharmonic on U. Consequently, we have

$$\begin{aligned}\log\|f(z_0)\| &= \log|\phi(f(z_0))| \\ &\leq \frac{1}{2\pi}\int_0^{2\pi} \log|\phi(f(z_0+re^{it}))|\,dt \\ &\leq \frac{1}{2\pi}\int_0^{2\pi} \log\|(f(z_0+re^{it}))\|\,dt,\end{aligned}$$

so $\log\|f\|$ does indeed satisfy the submean inequality. $\square$

Our first main theorem is an analogue of this result for the spectral radius.

Theorem 6.4.2 (Vesentini's Theorem) *If $f: U \to A$ is a holomorphic function from an open subset U of $\mathbb{C}$ to a Banach algebra A, then $\log \rho(f)$ is a subharmonic function on U.*

Proof. For $n \geq 1$ set

$$u_n(z) = 2^{-n}\log\|f(z)^{2^n}\| \quad (z \in U).$$

Since f^{2^n} is a holomorphic function, Lemma 6.4.1 implies that u_n is subharmonic on U. Also, because $\|a^{2^{n+1}}\| \leq \|a^{2^n}\|\|a^{2^n}\|$ for all $a \in A$, the sequence (u_n) is decreasing, and by the spectral radius formula (6.13) it converges to $\log\rho(f)$. Hence by Theorem 2.4.6 $\log\rho(f)$ is subharmonic on U. $\square$

The proof of Vesentini's theorem is deceptively simple. It belies the fact that, in many ways, the spectral radius function is quite badly behaved. For example, unlike the norm, it is not in general subadditive, nor submultiplicative, nor even continuous. Indeed, one can construct operators a and b on a Hilbert space such that $z \to \rho(az+b)$ is discontinuous almost everywhere! Thus subharmonicity is sometimes the only tool left available. Here is a typical application.

Corollary 6.4.3 *Let H be a real subspace of a Banach algebra A. If $\rho(h) = 0$ for all $h \in H$, then $\rho(h+ik) = 0$ for all $h, k \in H$.*

Proof. Note that we cannot simply say $\rho(h+ik) \leq \rho(h) + \rho(k)$, because ρ need not be subadditive. Instead we argue as follows. Let $h, k \in H$. The function $u(z) := \log\rho(h+zk)$ is subharmonic on $\mathbb{C}$ by Vesentini's theorem. Also, if $z \in \mathbb{R}$, then $h + zk \in H$, so $\rho(h+zk) = 0$, and consequently $u(z) = -\infty$. Now a subharmonic function on $\mathbb{C}$ which is $-\infty$ on $\mathbb{R}$ must be identically $-\infty$ on $\mathbb{C}$, because $\mathbb{R}$ is non-polar. Therefore $u \equiv -\infty$, and in particular $u(i) = -\infty$. Hence $\rho(h+ik) = 0$, as we wanted. $\square$

6.4. BANACH ALGEBRAS

A more sophisticated application is to prove the so-called uniqueness-of-norm theorem. In order to state this, we need a little more terminology.

If A is a Banach algebra with identity, then its *radical*, denoted by $\operatorname{rad}(A)$, is the intersection of all maximal left ideals in A. If A has no identity, then we define $\operatorname{rad}(A) = \operatorname{rad}(A^\#)$ (note that in this case, since A is itself a maximal left ideal in $A^\#$, we still have $\operatorname{rad}(A) \subset A$). It can be shown that $\operatorname{rad}(A)$ is always a closed two-sided ideal in A, and that $\rho(a) = 0$ for all $a \in \operatorname{rad}(A)$. Finally, A is called *semisimple* if $\operatorname{rad}(A) = \{0\}$.

Theorem 6.4.4 (Johnson's Theorem) *Let A and B be Banach algebras, and suppose that B is semisimple. Then every surjective homomorphism $\theta: A \to B$ is automatically continuous.*

Applying this to the identity map, we immediately deduce the following corollary, which shows that the algebraic and topological structures on a Banach algebra are tied together much more closely than one might suspect from the original definition.

Corollary 6.4.5 (Uniqueness-of-norm Theorem) *If A is a semisimple Banach algebra, then any other Banach-algebra norm on A is necessarily equivalent to the given one.* □

For instance, all the examples of Banach algebras listed at the beginning of this section are semisimple. Thus, up to equivalence, their given norms are the only ones we could have chosen.

On the other hand, Corollary 6.4.5 breaks down if A is not semisimple. As an extreme example, take any two inequivalent Banach-space norms on a suitable vector space A, and turn them into Banach-algebra norms simply by defining the product of every pair of elements of A to be zero. In this case $\operatorname{rad}(A) = A$, but it is also possible to construct examples where $\operatorname{rad}(A)$ is only one-dimensional.

Proof of Theorem 6.4.4. If A has an identity e_A then, since θ is a surjective homomorphism, $\theta(e_A)$ is necessarily an identity for B. If A has no identity, then we may adjoin one, e_A, and extend θ linearly to $A^\#$ by defining $\theta(e_A) = e_B$, where e_B is the identity of B if one exists, or one adjoined to B if not. The map θ, so extended, is still a surjective homomorphism. Thus we may as well suppose from the outset that A and B have identities e_A, e_B respectively, and that $\theta(e_A) = e_B$. Then θ maps invertible elements of A to invertible elements of B, and hence $\sigma(\theta(a)) \subset \sigma(a)$ for all $a \in A$. In particular,

(6.14) $$\rho(\theta(a)) \leq \rho(a) \quad (a \in A),$$

a property of θ that we shall exploit in the course of the proof.

Let (a_n) be a sequence in A such that $a_n \to 0$ in A and $\theta(a_n) \to b$ in B. By the closed graph theorem, it will follow that θ is continuous if we can show that $b = 0$. This, then, is the aim of the rest of the proof.

The first step, which is the key, is to show that $\rho(b) = 0$. It is tempting to deduce this directly from the fact that $\rho(\theta(a_n)) \leq \rho(a_n) \leq \|a_n\| \to 0$, but unfortunately we do not know that $\rho(\theta(a_n)) \to \rho(b)$, since ρ need not be continuous on B. Instead, a more subtle argument is needed. For $n \geq 1$, define $q_n \colon \mathbb{C} \to B$ by

$$q_n(z) = z\theta(a_n) + b - \theta(a_n).$$

By Vesentini's theorem, $\log \rho(q_n)$ is subharmonic on $\mathbb{C}$, and so by Theorem 2.6.8, $\max_{|z|=r} \log \rho(q_n(z))$ is a convex function of $\log r$. In particular, this implies that, for each $r > 0$,

(6.15) $$\rho(q_n(1))^2 \leq \max_{|z|=r} \rho(q_n(z)) \max_{|z|=1/r} \rho(q_n(z)).$$

Now certainly

$$\rho(q_n(z)) \leq \|q_n(z)\| \leq |z|\|\theta(a_n)\| + \|b - \theta(a_n)\|.$$

Also, if we choose $a \in A$ such that $\theta(a) = b$, then $q_n(z) = \theta(za_n + a - a_n)$, so by (6.14)

$$\rho(q_n(z)) \leq \rho(za_n + a - a_n) \leq |z|\|a_n\| + \|a - a_n\|.$$

Inserting these estimates into (6.15), we obtain

$$\rho(b)^2 \leq \bigl(r\|\theta(a_n)\| + \|b - \theta(a_n)\|\bigr)\bigl(r^{-1}\|a_n\| + \|a - a_n\|\bigr).$$

Letting $n \to \infty$, we get

$$\rho(b)^2 \leq r\|b\|\|a\|.$$

Lastly, letting $r \to 0$, we conclude that $\rho(b) = 0$, as claimed.

The next step is to show that $\rho(b'b) = 0$ for all $b' \in B$. Given $b' \in B$, choose $a' \in A$ with $\theta(a') = b'$. Then $a'a_n \to 0$ in A and $\theta(a'a_n) \to b'b$ in B, so, repeating the argument of the previous paragraph, we get $\rho(b'b) = 0$.

The third step is to show that $b \in \mathrm{rad}(B)$. Suppose, for a contradiction, that $b \notin \mathrm{rad}(B)$, so that b lies outside some maximal left ideal L. Then $Bb + L$ is a left ideal properly containing L, so it equals the whole of B, and in particular it contains the identity e_B. Thus there exists $b' \in B$ such that $e_B - b'b \in L$. But then $e_B - b'b$ cannot be invertible, which contradicts what we have already proved, namely that $\rho(b'b) = 0$. Therefore indeed $b \in \mathrm{rad}(B)$.

Finally, since B is semisimple, $\mathrm{rad}(B) = \{0\}$, so $b = 0$, and the proof is complete. $\square$

6.4. BANACH ALGEBRAS

If A is a *commutative* Banach algebra, namely one for which $ab = ba$ for all $a, b \in A$, then the spectral radius ρ *is* subadditive, submultiplicative and continuous. This is a simple consequence of the Gelfand representation theory, to be described in a moment. Thus, although Vesentini's theorem is of course still true, it is no longer so useful in this context because more powerful tools are available. On the other hand, there is now another subharmonicity theorem with interesting applications. To state this, we first need to give a brief outline of the Gelfand theory.

Let A be a commutative Banach algebra with identity e. A *character* on A is a homomorphism $\chi : A \to \mathbb{C}$ such that $\chi(e) = 1$. Every character is automatically a continuous linear functional of norm 1. The set of all characters, denoted by Φ_A, is a non-empty weak*-compact subset of the dual space of A. Given $f \in A$, we define $\widehat{f} : \Phi_A \to \mathbb{C}$ by $\widehat{f}(\chi) = \chi(f)$. This makes $\widehat{f}$ a continuous function on Φ_A, called the *Gelfand transform* of f. It can be shown that

$$\sigma(f) = \{\chi(f) : \chi \in \Phi_A\} = \widehat{f}(\Phi_A),$$

and therefore that

$$\rho(f) = \|\widehat{f}\|_{\Phi_A},$$

from which it follows easily that ρ is a continuous algebra seminorm on A, as claimed above.

A *boundary* for A is a subset S of Φ_A such that $\|\widehat{f}\|_{\Phi_A} = \|\widehat{f}\|_S$ for all $f \in A$. It can be shown that the intersection of all closed boundaries is again a closed boundary, necessarily the minimal one. It is called the *Shilov boundary*, denoted by ∂_A. Note that if $f \in A$ and $z_0 \in \widehat{f}(\Phi_A)$, say $z_0 = \widehat{f}(\chi_0)$, then for all $z \in \mathbb{C} \setminus \widehat{f}(\Phi_A)$ we have

$$\frac{1}{|z - z_0|} = |\chi_0(ze - f)^{-1}| \leq \sup_{\chi \in \partial_A} |\chi(ze - f)^{-1}| = \frac{1}{\operatorname{dist}(z, \widehat{f}(\partial_A))}.$$

It follows that

$$\partial \widehat{f}(\Phi_A) \subset \widehat{f}(\partial_A),$$

or equivalently, that $\widehat{f}(\Phi_A) \setminus \widehat{f}(\partial_A)$ is an open (possibly empty) subset of $\mathbb{C}$.

For further information about the Gelfand representation and the Shilov boundary, see for instance [30, 58, 63]. We are now ready to state our second subharmonicity theorem.

Theorem 6.4.6 (Wermer's Theorem) *Let A be a commutative Banach algebra with identity, and let $f, g \in A$. If we define $\rho_{f,g} \colon \widehat{g}(\Phi_A) \to \mathbb{C}$ by*

$$\rho_{f,g}(z) = \sup\{|\widehat{f}(\chi)| : \chi \in \Phi_A, \widehat{g}(\chi) = z\},$$

then $\log \rho_{f,g}$ is upper semicontinuous on the compact set $\widehat{g}(\Phi_A)$, and subharmonic on the open set $\widehat{g}(\Phi_A) \setminus \widehat{g}(\partial_A)$.

The proof relies on a lemma, which also explains the notation $\rho_{f,g}$.

Lemma 6.4.7 *Let A, f, g be as in the Theorem.*

(a) *If $z \in \widehat{g}(\Phi_A)$, then*
$$\rho_{f,g}(z) = \rho_{A/I}([f]),$$
where I is the closed ideal $\overline{(g - ze)A}$, and $\rho_{A/I}$ denotes the spectral radius in the quotient Banach algebra A/I.

(b) *If, further, $\phi \colon A \to \mathbb{C}$ is a continuous linear functional which vanishes on I, then*
$$\limsup_{n \to \infty} |\phi(f^n)|^{1/n} \leq \rho_{f,g}(z).$$

Proof. (a) Let $\pi \colon A \to A/I$ denote the quotient map. Then $\xi \mapsto \xi \circ \pi$ is a bijection between the characters on A/I and those characters on A which vanish on I. Moreover, a character χ on A vanishes on I if and only if $\chi(g) = z$. Thus

$$\begin{aligned}
\rho_{A/I}([f]) &= \sup\{|\xi \circ \pi(f)| : \xi \in \Phi_{A/I}\} \\
&= \sup\{|\chi(f)| : \chi \in \Phi_A, \chi(I) = \{0\}\} \\
&= \sup\{|\chi(f)| : \chi \in \Phi_A, \chi(g) = z\} \\
&= \rho_{f,g}(z).
\end{aligned}$$

(b) If $\phi(I) = \{0\}$, then it follows that

$$|\phi(h)| \leq \|\phi\| \, \|[h]\|_{A/I} \quad (h \in A).$$

Hence, by the spectral radius formula,

$$\limsup_{n \to \infty} |\phi(f^n)|^{1/n} \leq \limsup_{n \to \infty} \|\phi\|^{1/n} \|[f]^n\|_{A/I}^{1/n} = \rho_{A/I}([f]),$$

and combining this with part (a) gives the result. $\square$

6.4. BANACH ALGEBRAS

Proof of Theorem 6.4.6. We first show that $\log \rho_{f,g}$ is upper semicontinuous on $\widehat{g}(\Phi_A)$. Given $\alpha \in \mathbb{R}$, we have

$$\{z \in \widehat{g}(\Phi_A) : \log \rho_{f,g}(z) \geq \alpha\} = \widehat{g}(\Psi),$$

where

$$\Psi = \{\chi \in \Phi_A : \log |\widehat{f}(\chi)| \geq \alpha\}.$$

As Φ_A is compact, so too is Ψ, and hence also $\widehat{g}(\Psi)$. It follows that the set $\{\log \rho_{f,g} < \alpha\}$ is open in $\widehat{g}(\Phi_A)$, which proves upper semicontinuity.

To show that $\log \rho_{f,g}$ is subharmonic on $\widehat{g}(\Phi_A) \setminus \widehat{g}(\partial_A)$, we shall verify that it satisfies the submean inequality there. Fix $z_0 \in \widehat{g}(\Phi_A) \setminus \widehat{g}(\partial_A)$, and choose $\chi_0 \in \Phi_A$ such that $\widehat{g}(\chi_0) = z_0$ and $|\widehat{f}(\chi_0)| = \rho_{f,g}(z_0)$. Since ∂_A is a boundary, $|\widehat{h}(\chi_0)| \leq \|\widehat{h}\|_{\partial_A}$ for all $h \in A$, so the map $\psi : \widehat{h} \mapsto \widehat{h}(\chi_0)$ may be regarded as a linear functional of norm 1 on a subspace of $C(\partial_A)$. By the Hahn–Banach theorem, ψ can be extended to the whole of $C(\partial_A)$ without increasing its norm. For $z \in \widehat{g}(\Phi_A) \setminus \widehat{g}(\partial_A)$, define $\phi_z : A \to \mathbb{C}$ by

$$\phi_z(h) = \psi\left(\frac{\widehat{g} - z_0}{\widehat{g} - z} \widehat{h}\right) \quad (h \in A).$$

Then ϕ_z is a continuous linear functional on A since

$$|\phi_z(h)| \leq \left\|\frac{\widehat{g} - z_0}{\widehat{g} - z} \widehat{h}\right\|_{\partial_A} \leq \left\|\frac{\widehat{g} - z_0}{\widehat{g} - z}\right\|_{\partial_A} \|h\| \quad (h \in A).$$

Also,

$$\phi_z\big((g - ze)h\big) = \psi\big((\widehat{g} - z_0)\widehat{h}\big) = (\widehat{g}(\chi_0) - z_0)\widehat{h}(\chi_0) = 0 \quad (h \in A),$$

so ϕ_z vanishes on $\overline{(g - ze)A}$, and consequently by Lemma 6.4.7,

$$(6.16) \qquad \limsup_{n \to \infty} |\phi_z(f^n)|^{1/n} \leq \rho_{f,g}(z) \quad (z \in \widehat{g}(\Phi_A) \setminus \widehat{g}(\partial_A)).$$

Moreover, if $z = z_0$, then

$$\phi_{z_0}(h) = \psi(\widehat{h}) = \widehat{h}(\chi_0) \quad (h \in A),$$

and so in particular,

$$(6.17) \qquad \limsup_{n \to \infty} |\phi_{z_0}(f^n)|^{1/n} = |\widehat{f}(\chi_0)| = \rho_{f,g}(z_0).$$

Now for each $h \in A$, the function $z \mapsto \phi_z(h)$ is holomorphic on the set $\widehat{g}(\Phi_A) \setminus \widehat{g}(\partial_A)$. Hence if $r < \mathrm{dist}(z_0, \widehat{g}(\partial_A))$, then

$$\log |\phi_{z_0}(h)| \leq \frac{1}{2\pi} \int_0^{2\pi} \log |\phi_{z_0 + re^{it}}(h)| \, dt.$$

Setting $h = f^n$ and dividing by n, we get

$$\log |\phi_{z_0}(f^n)|^{1/n} \leq \frac{1}{2\pi} \int_0^{2\pi} \log |\phi_{z_0 + re^{it}}(f^n)|^{1/n} \, dt.$$

Taking $\limsup_{n \to \infty}$ of both sides, and applying Fatou's lemma and the estimates (6.16) and (6.17), we obtain

$$\log \rho_{f,g}(z_0) \leq \frac{1}{2\pi} \int_0^{2\pi} \log \rho_{f,g}(z_0 + re^{it}) \, dt.$$

This proves the submean inequality for $\log \rho_{f,g}$, and completes the proof of the theorem. $\square$

As an application, we prove a converse to the maximum modulus principle for holomorphic functions.

Theorem 6.4.8 (Rudin's Theorem) *Let A be an algebra of continuous functions on a closed disc $\overline{\Delta}$ in $\mathbb{C}$. Suppose that A contains the function $j(z) \equiv z$, and that every $f \in A$ satisfies*

(6.18) $$\|f\|_{\overline{\Delta}} = \|f\|_{\partial \Delta}.$$

Then every $f \in A$ is holomorphic on Δ.

Proof. It is not assumed that A contains the constant function $e(z) \equiv 1$. However, if $f \in A$ and $\alpha \in \mathbb{C}$, then the functions $(f + \alpha e)^n j$ belong to A for all $n \geq 1$, so

$$\|(f + \alpha e)^n j\|_{\overline{\Delta}} = \|(f + \alpha e)^n j\|_{\partial \Delta} \leq \|f + \alpha e\|_{\partial \Delta}^n \|j\|_{\partial \Delta}.$$

Therefore, taking n-th roots and letting $n \to \infty$, it follows that

$$\|f + \alpha e\|_{\overline{\Delta}} = \|f + \alpha e\|_{\partial \Delta},$$

and so we might as well adjoin e to A. Also (6.18) clearly still holds if f lies in the uniform closure of A. Thus, replacing A by its closure, we can suppose that $(A, \|\cdot\|_{\overline{\Delta}})$ is a commutative Banach algebra with identity, so that all the preceding theory applies.

As evaluation at each point of $\overline{\Delta}$ is a character on A, we have $\widehat{j}(\Phi_A) \supset \overline{\Delta}$. Also, (6.18) implies that the set of evaluations at points of $\partial \Delta$ forms a closed boundary for A, and so $\widehat{j}(\partial_A) \subset \partial \Delta$. Hence, because of the general inclusion $\partial \widehat{j}(\Phi_A) \subset \widehat{j}(\partial_A)$, it follows that $\widehat{j}(\Phi_A) = \overline{\Delta}$ and $\widehat{j}(\partial_A) = \partial \Delta$. (In fact the proof will show that $\widehat{j}$ is a homeomorphism of Φ_A onto $\overline{\Delta}$, but we do not know this *a priori*.)

6.4. BANACH ALGEBRAS

The next step is to show that if χ is a character with $\hat{j}(\chi) = z_0 \in \partial\Delta$, then $\hat{f}(\chi) = f(z_0)$ for all $f \in A$. To prove this, choose a sequence of polynomials (q_n) such that $q_n(z_0) = 1$ for all n, and $q_n \to 0$ boundedly and locally uniformly on $\overline{\Delta} \setminus \{z_0\}$. For example, writing $\overline{\Delta} = \overline{\Delta}(w, r)$, we could take

$$q_n(z) = \frac{1}{2^n}\left(1 + \frac{z-w}{z_0-w}\right)^n.$$

Then, given $f \in A$, the sup-norms $\|(f - f(z_0))q_n\|_{\overline{\Delta}} \to 0$, and since χ is a continuous linear functional, it follows that $\chi((f - f(z_0))q_n) \to 0$. But also, because q_n is a polynomial, we have $\chi(q_n) = q_n(\chi(j)) = q_n(z_0) = 1$ for all n. Consequently $\chi(f - f(z_0)) = 0$, or in other words $\hat{f}(\chi) = f(z_0)$, as claimed.

Now fix $f \in A$, and define $u, v: \overline{\Delta} \to \mathbb{R}$ by

$$u(z) = \sup\{\operatorname{Re}\hat{f}(\chi) : \chi \in \Phi_A, \hat{j}(\chi) = z\},$$
$$v(z) = \inf\{\operatorname{Re}\hat{f}(\chi) : \chi \in \Phi_A, \hat{j}(\chi) = z\}.$$

If $z \in \overline{\Delta}$, then evaluation at z is a character χ such that $\hat{j}(\chi) = z$, so

(6.19) $$v(z) \leq \operatorname{Re} f(z) \leq u(z) \quad (z \in \overline{\Delta}),$$

and if $z \in \partial\Delta$, then we have seen that it is the only such character, whence

(6.20) $$v(z) = \operatorname{Re} f(z) = u(z) \quad (z \in \partial\Delta).$$

Now in the notation of Theorem 6.4.6, $u(z) = \log \rho_{e^f, j}(z)$, and so by that theorem u is upper semicontinuous on $\overline{\Delta}$ and subharmonic on Δ. Likewise, v is lower semicontinuous on $\overline{\Delta}$ and superharmonic on Δ. Also (6.20) implies that $u - v = 0$ on $\partial\Delta$. Hence by the maximum principle $u - v \leq 0$ on Δ. Combining this with (6.19), we see that $u = v = \operatorname{Re} f$ on Δ. In particular, $\operatorname{Re} f$ is both subharmonic and superharmonic on Δ, and therefore harmonic there. Similarly $\operatorname{Im} f$ is harmonic, and so f satisfies Laplace's equation $\Delta f = 0$. Repeating the argument with f replaced by jf, we get $\Delta(jf) = 0$. Expanding this and simplifying, we find that f satisfies the Cauchy–Riemann equations on Δ, and hence is holomorphic there. □

This proof has the advantage, over some others, that it generalizes easily to other types of compact set. The only modification needed is in the construction of the polynomials q_n, for which a result like the Walsh–Lebesgue theorem (Theorem 6.3.4) can be used. More about this can be found in [63, §25].

A Banach algebra A is called *radical* if $\mathrm{rad}(A) = A$. Thus radical is the opposite extreme of semisimple, though, as we shall see, radical Banach algebras may still be used to analyse semisimple ones. It can be shown that a Banach algebra is radical if and only if the spectral radius of every element is zero, and so in this case the spectral-theoretic techniques developed above are of little help, and other tools are needed.

One such tool is the notion of a holomorphic semigroup. Let A be an arbitrary Banach algebra, and let H denote the half-plane $\{z : \mathrm{Re}\, z > 0\}$. A map $z \mapsto a(z) \colon H \to A$ is a *holomorphic semigroup* if it is holomorphic and satisfies $a(z+w) = a(z)a(w)$ for all $z, w \in H$. It is customary to write $a(z)$ as a^z, so the semigroup law takes the more suggestive form:

$$a^{z+w} = a^z a^w \quad (z, w \in H).$$

Also we write a^1 simply as a.

Notice that no assumption has been made regarding the behaviour of a^z as $z \to 0$. However, if $b \in aA$, say $b = ac$, then

$$a^{1/n} b = a^{1 + 1/n} c \to a^1 c = b \quad \text{as } n \to \infty.$$

It follows that if $\overline{aA} = \overline{Aa} = A$ and $\sup_{x > 0} \|a^x\| < \infty$, then $(a^{1/n})_{n \geq 1}$ is a *bounded approximate identity* for A, i.e. a bounded sequence $(e_n)_{n \geq 1}$ in A such that

$$\lim_{n \to \infty} e_n b = \lim_{n \to \infty} b e_n = b \quad \text{for all } b \in A.$$

Conversely, it can be shown that if A has a bounded approximate identity, then there is a holomorphic semigroup $z \mapsto a^z \colon H \to A$ such that $\overline{aA} = \overline{Aa} = A$ and $\sup_{x > 0} \|a^x\| < \infty$. In particular, this applies to many radical Banach algebras, for even though no such algebra can have an identity (the spectral radius would have to be zero), it often does have a bounded approximate identity.

The classic reference for holomorphic semigroups is Hille–Phillips [37]. For more information regarding connections with bounded approximate identities, see Sinclair's book [61].

One reason that holomorphic semigroups can be useful in studying a radical Banach algebra A is that every such semigroup (a^z) in A is automatically subject to severe growth constraints. For example, since $\rho(a^z) = 0$ for each $z \in H$, it follows from the spectral radius formula and continuity that

(6.21) $$\lim_{t \to \infty} \|a^{tz}\|^{1/t} = 0 \quad (z \in H).$$

Perhaps more surprisingly, the semigroup actually has to grow larger along vertical lines. This is made precise in our third main theorem of the section.

6.4. BANACH ALGEBRAS

Theorem 6.4.9 *Let A be a radical Banach algebra, and let $z \mapsto a^z \colon H \to A$ be a holomorphic semigroup such that $a^z \not\equiv 0$. Then for each $x > 0$,*

$$\int_{-\infty}^{\infty} \frac{\log^+ \|a^{x+iy}\|}{1+y^2}\, dy = \infty.$$

Proof. By scaling, it is enough to prove the result for $x = 1$. Suppose, then, for a contradiction, that

$$\int_{-\infty}^{\infty} \frac{\log^+ \|a^{1+it}\|}{1+t^2}\, dt = I < \infty.$$

We begin by establishing a bound, albeit a weak one, for $\|a^{3+z}\|$. Put

$$E = \left\{ t \in \mathbb{R} : \frac{\log^+ \|a^{1+it}\|}{1+t^2} > 1/2 \right\},$$

so that E is an open set of Lebesgue measure at most $2I$. Then, given $y \in \mathbb{R}$, there exists $t \in \mathbb{R} \setminus E$ with $|y - t| \leq I$, and so

$$\begin{aligned}
\|a^{2+iy}\| &\leq \|a^{1+it}\|\, \|a^{1+i(y-t)}\| \\
&\leq e^{1/2 + t^2/2} \|a^{1+i(y-t)}\| \\
&\leq e^{1/2 + I^2 + y^2} \|a^{1+i(y-t)}\| \\
&\leq C_1 e^{y^2},
\end{aligned}$$

where $C_1 = e^{1/2 + I^2} \sup_{|s| \leq I} \|a^{1+is}\|$. Also, equation (6.21) implies that $\|a^{1+x}\|$ is bounded for $x > 0$. Hence, writing $z = x + iy$, we have

(6.22) $\qquad \|a^{3+z}\| \leq \|a^{1+x}\|\, \|a^{2+iy}\| \leq C_2 e^{y^2} \quad (z \in H),$

where $C_2 = C_1 \sup_{x > 0} \|a^{1+x}\|$.

Next, we introduce the function $h \colon H \to [0, \infty]$ defined by

$$h(z) = \frac{1}{\pi} \int_{-\infty}^{\infty} \frac{\operatorname{Re} z}{|z - it|^2} \log^+ \|a^{3+it}\|\, dt \quad (z \in H).$$

We claim that h is a positive harmonic function on H satisfying

(6.23) $\qquad \liminf_{z \to it} h(z) \geq \log^+ \|a^{3+it}\| \quad (it \in \partial H \setminus \{\infty\}).$

To see this, take an increasing sequence of continuous, compactly supported functions $\phi_n \colon \partial H \setminus \{\infty\} \to \mathbb{R}$, such that $\phi_n(it) \uparrow \log^+ \|a^{3+it}\|$ for each t.

By (a rotated version of) Theorem 4.3.13, their Poisson integrals on H are given by

$$P_H\phi_n(z) = \frac{1}{\pi}\int_{-\infty}^{\infty} \frac{\operatorname{Re} z}{|z-it|^2}\phi_n(it)\, dt \quad (z \in H).$$

Therefore, by the monotone convergence theorem, $P_H\phi_n$ increases pointwise to h, and so by Harnack's theorem, either h is harmonic on H or $h \equiv \infty$. Since

$$h(1) = \frac{1}{\pi}\int_{-\infty}^{\infty} \frac{\log^+ \|a^{3+it}\|}{1+t^2}\, dt \leq \log^+ \|a^2\| + \frac{I}{\pi} < \infty,$$

it follows that h is in fact harmonic. Clearly it is also positive. Finally, for each n we have

$$\liminf_{z\to it} h(z) \geq \liminf_{z\to it} P_H\phi_n(z) = \phi_n(it) \qquad (it \in \partial H \setminus \{\infty\}),$$

and so letting $n \to \infty$ we get (6.23), as claimed.

We are now ready for the *dénouement*. Take $M > 0$ and define

$$u(z) = \log\|a^{3+z}\| - h(z) + M\operatorname{Re} z \quad (z \in H).$$

Then u is subharmonic on H, by Lemma 6.4.1. Also, from (6.23) it follows that

$$\limsup_{z\to it} u(z) \leq 0 \quad (it \in \partial H \setminus \{\infty\}).$$

Furthermore, using (6.22) we have

$$u(x+iy) \leq (\log C_2 + y^2) + 0 + Mx \leq \log C_2 + y^2 + M^2 + x^2,$$

and hence

$$u(z) \leq C_3 + |z|^2 \quad (z \in H),$$

where $C_3 = \log C_2 + M^2$. Lastly, equation (6.21) implies that u is bounded above on the ray $\arg z = \theta$ for each $\theta \in (-\pi/2, \pi/2)$, and in particular for $\theta = \pm\pi/6$. We can therefore apply the Phragmén-Lindelöf theorem (Theorem 2.3.7) with $\gamma = 3$ on each of the three sectors $\{-\pi/2 < \arg z < -\pi/6\}$, $\{-\pi/6 < \arg z < \pi/6\}$ and $\{\pi/6 < \arg z < \pi/2\}$, to deduce that u is bounded above on each sector, and hence on H. Consequently, we may use Theorem 2.3.7 once again, this time with $\gamma = 1$, to conclude that $u \leq 0$ on H. Hence

$$\log\|a^{3+z}\| \leq h(z) - M\operatorname{Re} z \quad (z \in H).$$

Since M is arbitrary, it follows that $\log\|a^{3+z}\| = -\infty$ for all $z \in H$. In other words, $a^z = 0$ for $\operatorname{Re} z > 3$, and as $z \mapsto a^z$ is holomorphic, this implies that $a^z \equiv 0$ on H. We have thus arrived at the desired contradiction. $\square$

6.4. BANACH ALGEBRAS

We are now going to use this result to prove Wiener's general tauberian theorem. In its classical form, this theorem says that if $\phi \in L^\infty(\mathbf{R})$ and
$$\lim_{t \to \infty} f_0 * \phi(t) = 0,$$
where $f_0 \in L^1(\mathbf{R})$ is a function whose Fourier transform satisfies $\widehat{f_0}(\tau) \neq 0$ for all $\tau \in \mathbf{R}$, then in fact

(6.24) $$\lim_{t \to \infty} f * \phi(t) = 0$$

for every $f \in L^1(\mathbf{R})$.

It is convenient first to reformulate this in terms of Banach algebras. If we regard $L^1(\mathbf{R})$ as a commutative Banach algebra, with multiplication given by convolution, then the set I of all $f \in L^1(\mathbf{R})$ which satisfy (6.24) is a closed ideal containing f_0. It therefore suffices to show that the only such ideal is $L^1(\mathbf{R})$ itself, and in fact we shall prove a slightly stronger result.

Theorem 6.4.10 (Wiener's Tauberian Theorem) *Let I be a closed ideal in $L^1(\mathbf{R})$, and suppose that for each $\tau \in \mathbf{R}$ there exists $f \in I$ with $\widehat{f}(\tau) \neq 0$. Then $I = L^1(\mathbf{R})$.*

Before beginning the proof, it is perhaps worth remarking that what makes this result more difficult is the fact that $L^1(\mathbf{R})$ has no identity—if it did, we could argue that every proper ideal is contained in a maximal ideal, which is the kernel of some character $f \mapsto \widehat{f}(\tau)$. The idea of using holomorphic semigroups to overcome this problem is due to Esterle [28].

Proof. Let A be the quotient Banach algebra $L^1(\mathbf{R})/I$, and denote by $\pi: L^1(\mathbf{R}) \to A$ the quotient map. Let χ be a character on $A^\#$, the algebra obtained by adjoining an identity to A. Then $\chi \circ \pi: L^1(\mathbf{R}) \to \mathbf{C}$ is a homomorphism which vanishes on I. Now a non-zero homomorphism from $L^1(\mathbf{R})$ to $\mathbf{C}$ has the form $f \mapsto \widehat{f}(\tau)$ for some $\tau \in \mathbf{R}$, so the hypothesis on I prevents it from vanishing on I. Therefore $\chi \circ \pi$ must be the zero homomorphism, and consequently $\chi = 0$ on A. By the Gelfand theory, it follows that every element of A has spectral radius zero, and thus A is a radical Banach algebra.

Next, for $z \in H$, define $g^z \in L^1(\mathbf{R})$ by
$$g^z(t) = \frac{1}{\sqrt{4\pi z}} e^{-t^2/4z} \quad (t \in \mathbf{R}).$$

By considering the Fourier transforms of both sides, we have
$$g^{z+w} = g^z * g^w \quad (z, w \in H),$$

so (g^z) is a holomorphic semigroup in $L^1(\mathbb{R})$ (the so-called *Gauss semigroup*). Moreover, a simple computation shows that

$$\|g^{1+iy}\|_1 = (1+y^2)^{1/4} \quad (y \in \mathbb{R}).$$

Hence if we define $a^z = \pi(g^z)$, then (a^z) is a holomorphic semigroup in A such that

$$\int_{-\infty}^{\infty} \frac{\log^+ \|a^{1+iy}\|}{1+y^2}\, dy < \infty.$$

Since A is radical, Theorem 6.4.9 implies that $a^z \equiv 0$, in other words,

$$g^z \in I \quad (z \in H).$$

Now take an arbitrary function $f \in L^1(\mathbb{R})$. Since $g^{1/n} \in I$ and I is an ideal, the product $g^{1/n} * f \in I$ for each n. Also, a calculation similar to that in the proof of Theorem 1.2.4 (b) shows that $\|g^{1/n} * f - f\|_1 \to 0$ as $n \to \infty$. As I is closed, we conclude that $f \in I$. This proves that $I = L^1(\mathbb{R})$. □

6.5 Complex Dynamics

Let $q(z) = \sum_{j=0}^{d} a_j z^j$ be a polynomial of degree d. We are going to study the *dynamics* of q, namely the behaviour of the iterates $q^n = q \circ \cdots \circ q$ as $n \to \infty$. If $d = 0$ or 1 then this is relatively uninteresting because one can easily write down an explicit formula for q^n, so we shall assume from the outset that we are in the more interesting case $d \geq 2$.

The *attracting basin of* ∞ of q is the set

$$F_\infty := \{z \in \mathbb{C}_\infty : q^n(z) \to \infty \text{ as } n \to \infty\}.$$

Since $d \geq 2$, it follows that $|q(z)| \geq 2|z|$ for all sufficiently large $|z|$, so F_∞ contains a connected open neighbourhood U of ∞ such that $U \subset q^{-1}(U)$. Since $F_\infty = \cup_n q^{-n}(U)$, it follows that F_∞ itself is a connected open set, and that $q^n \to \infty$ locally uniformly on F_∞. Also F_∞ is *completely invariant*, i.e. $q^{-1}(F_\infty) = F_\infty$.

The *Julia set* of q is defined by

$$J := \partial F_\infty.$$

It is therefore a compact subset of $\mathbb{C}$, and it too is completely invariant: $q^{-1}(J) = J$. It turns out that the Julia set plays a fundamental rôle in the dynamics of q.

6.5. COMPLEX DYNAMICS

Here are two simple examples. If $q(z) = z^2$, then $q^n(z) = z^{2^n}$, and so

$$F_\infty = \{z : |z| > 1\} \quad \text{and} \quad J = \{z : |z| = 1\}.$$

If $q(z) = z^2 - 2$, then a simple calculation shows that $q^n(w + 1/w) = w^{2^n} + 1/w^{2^n}$, from which it follows that

$$F_\infty = \mathbf{C}_\infty \setminus [-2, 2] \quad \text{and} \quad J = [-2, 2].$$

However, these examples are atypical. In general the Julia set has a very complicated structure. Indeed, it is probably fair to say that it is because the simple definitions above give rise to such complex sets, as witnessed by the remarkable computer-generated pictures now available, that the subject has become so popular in recent years. At the same time, the mathematics behind polynomial iteration is also very rich and beautiful—see the three recent books [13, 23, 62], for example.

In this section we take a look at some interactions between complex dynamics and potential theory. The first result is that the Julia set is always non-polar, and in fact we have the following quantitative statement.

Theorem 6.5.1 *If $q(z) = \sum_{j=0}^d a_j z^j$ is a polynomial of degree $d \geq 2$, then its Julia set has capacity $c(J) = 1/|a_d|^{1/(d-1)}$.*

Before beginning the proof, we introduce a technical device which will simplify several of the arguments in this section. Let $m(z) = \alpha z + \beta$ be a polynomial of degree 1, and let $\tilde{q}$ be the conjugate

$$\tilde{q} = m \circ q \circ m^{-1}.$$

Then $\tilde{q}$ is again a polynomial of degree d, and $\tilde{q}^n = m \circ q^n \circ m^{-1}$ for each n, so that $\tilde{q}$ has essentially the same dynamics as q. In particular, it follows that the attracting basin of ∞ and the Julia set of $\tilde{q}$ are given by $\tilde{F}_\infty = m(F_\infty)$ and $\tilde{J} = m(J)$ respectively. The advantage of conjugating q in this fashion is that $\tilde{q}$ can be made algebraically simpler than q by choosing α, β appropriately. For example, if $\alpha = a_d^{1/(d-1)}$ then $\tilde{q}$ is monic, and taking $\beta = a_{d-1}/d$ we can further ensure that $\tilde{q}(z) = z^d + O(z^{d-2})$ as $z \to \infty$.

Proof of Theorem 6.5.1. Suppose first that q is monic. Put $U = \{z \in \mathbf{C}_\infty : |z| > R\}$, where R is chosen large enough so that $|q(z)| \geq 2|z|$ ($z \in U$). For $n \geq 0$, define $K_n = \mathbf{C}_\infty \setminus q^{-n}(U)$. Then $K_n = q^{-1}(K_{n-1})$ for each n, so by Theorem 5.2.5

$$c(K_n) = c(K_{n-1})^{1/d} = \cdots = c(K_0)^{1/d^n}.$$

Also, since $U \subset q^{-1}(U)$, we see that $K_n \downarrow K$, where $K = \mathbb{C}_\infty \setminus \cup_n q^{-n}(U)$, so by Theorem 5.1.3

$$c(K) = \lim_{n\to\infty} c(K_n) = \lim_{n\to\infty} c(K_0)^{1/d^n} = 1.$$

Lastly, as $\cup_n q^{-n}(U) = F_\infty$, it follows that $\partial K = \partial F_\infty = J$, and so by Theorem 5.1.2 (d) we conclude that $c(J) = 1$.

Now take a general q, and let $m(z) = a_d^{1/(d-1)} z$, so that $\widetilde{q} := m \circ q \circ m^{-1}$ is monic. By what we have just proved, $c(\widetilde{J}) = 1$. Since $\widetilde{J} = m(J)$, it follows that $c(J) = 1/|a_d|^{1/(d-1)}$, as stated. $\square$

Corollary 6.5.2 *With q, J as above, the diameter of J satisfies*

$$\mathrm{diam}(J) \geq 2/|a_d|^{1/(d-1)}.$$

Moreover, if J is connected, then

$$\mathrm{diam}(J) \leq 4/|a_d|^{1/(d-1)}.$$

Proof. Combine Theorem 6.5.1 with Theorems 5.3.4 and 5.3.2 (a). $\square$

The examples of $q(z) = z^2$ and $q(z) = z^2 - 2$ show that both these inequalities are sharp. We could also apply Theorem 5.3.5, to obtain an upper bound for the area of J. However, at the time of writing it is actually still an open problem whether every Julia set has area zero!

As F_∞ has non-polar complement, it has a Green's function. Our next result gives an important invariance property of this function.

Theorem 6.5.3 *Let q be a polynomial of degree $d \geq 2$, and let F_∞ be the attracting basin of ∞ for q. Then*

$$g_{F_\infty}(q(z), \infty) = d\, g_{F_\infty}(z, \infty) \quad (z \in F_\infty).$$

Proof. Set $D = F_\infty \setminus \{\infty\}$, and define $h: D \to \mathbb{R}$ by

$$h(z) = g_{F_\infty}(q(z), \infty) - d\, g_{F_\infty}(z, \infty) \quad (z \in D).$$

Then h is harmonic on D. Also,

$$h(z) = \log|q(z)| - d\log|z| + O(1) = O(1) \quad \text{as } z \to \infty,$$

from which it follows that h is bounded on D. Finally, if I denotes the set of irregular boundary points of F_∞, then $I \cup q^{-1}(I) \cup \{\infty\}$ is a polar subset of ∂D, and

$$\lim_{z\to\zeta} h(z) = 0 \quad (\zeta \in \partial D \setminus (I \cup q^{-1}(I) \cup \{\infty\})).$$

Therefore by the extended maximum principle (Theorem 3.6.9) $h \equiv 0$ on D, which gives the result. $\square$

6.5. COMPLEX DYNAMICS

A nice consequence of this is that the Green's function has a purely dynamical interpretation.

Corollary 6.5.4 *With q, F_∞ as in the Theorem,*

$$g_{F_\infty}(z, \infty) = \lim_{n \to \infty} d^{-n} \log |q^n(z)| \quad (z \in F_\infty),$$

the convergence being uniform on compact subsets of F_∞.

Proof. From the definition of Green's function,

$$g_{F_\infty}(z, \infty) = \log |z| + O(1) \quad \text{as } z \to \infty.$$

Using Theorem 6.5.3, it follows that for $z \in F_\infty$,

$$g_{F_\infty}(z, \infty) = d^{-n} g_{F_\infty}(q^n(z), \infty) = d^{-n} \log |q^n(z)| + O(d^{-n}) \quad \text{as } n \to \infty,$$

the convergence being uniform on compact subsets of F_∞ because $q^n \to \infty$ locally uniformly there. □

This result gives an idea of the asymptotic behaviour of q^n on F_∞. In fact, a slightly more careful analysis shows that, if $q(z) = z^d + O(z^{d-2})$, then

$$|q^n(z)| = e^{d^n g_{F_\infty}(z,\infty)} + O(e^{-d^n g_{F_\infty}(z,\infty)}).$$

The example of $q(z) = z^2 - 2$ shows that the error term is sharp.

As another application of Theorem 6.5.3, we can prove that the Julia set has no isolated points, and indeed something much stronger.

Corollary 6.5.5 *The attracting basin F_∞ is a regular domain for the Dirichlet problem, and the Julia set J is non-thin at every point of itself.*

Proof. Since $g_{F_\infty}(z, \infty)$ is bounded near ∂F_∞, we can find a constant $C > 0$ such that the set

$$\{z \in F_\infty : g_{F_\infty}(z, \infty) \geq C\}$$

is compact. Using Theorem 6.5.3 and the complete invariance of F_∞, we then have

$$\{z \in F_\infty : g_{F_\infty}(z, \infty) \geq Cd^{-n}\} = q^{-n}(\{z \in F_\infty : g_{F_\infty}(z, \infty) \geq C\}),$$

so this set too is compact. It follows from this that if $\zeta \in \partial F_\infty$, then $\lim_{z \to \zeta} g_{F_\infty}(z, \infty) = 0$, and so by Theorem 4.4.9 ζ is a regular point. This shows that F_∞ is a regular domain for the Dirichlet problem.

Using Theorem 4.2.4, we can immediately deduce that $K := \mathbb{C}_\infty \setminus F_\infty$ is non-thin at each point of itself, and we want to prove that the same holds for $J = \partial K$. Suppose, for a contradiction, that there exists $\zeta \in J$ and a function u subharmonic on a neighbourhood of ζ such that

$$\limsup_{\substack{z \to \zeta \\ z \in J \setminus \{\zeta\}}} u(z) < u(\zeta). \tag{6.25}$$

By Exercise 3.8.3 or Theorem 5.4.2, there is a sequence of circles $\partial \Delta(\zeta, r_n)$ with $r_n \to 0$ such that $\partial \Delta(\zeta, r_n) \cap J = \varnothing$ for all n. Also each circle must intersect F_∞, otherwise it would disconnect F_∞. Hence $\partial \Delta(\zeta, r_n) \subset F_\infty$ for all n. It follows that every point of K close enough to ζ lies in a component of K all of which lies close to ζ, and hence by the maximum principle

$$\limsup_{\substack{z \to \zeta \\ z \in K \setminus \{\zeta\}}} u(z) = \limsup_{\substack{z \to \zeta \\ z \in J \setminus \{\zeta\}}} u(z).$$

Taken together with (6.25), this implies that K is thin at ζ, contradicting what we already know. Therefore J is non-thin at every point of itself. □

The harmonic measure ω_{F_∞} on $\partial F_\infty = J$ also exhibits an important invariance property.

Theorem 6.5.6 *Let q be a polynomial of degree $d \geq 2$, and let F_∞ and J be its attracting basin of ∞ and Julia set respectively. Then for each Borel subset B of J,*

$$\omega_{F_\infty}(q(z), B) = \omega_{F_\infty}(z, q^{-1}(B)) \quad (z \in F_\infty).$$

Proof. Given a continuous map $\phi \colon J \to \mathbb{R}$, both $(H_{F_\infty}\phi) \circ q$ and $H_{F_\infty}(\phi \circ q)$ are solutions to the Dirichlet problem on F_∞ with boundary data $\phi \circ q$, so by uniqueness they must be equal. From the definition of harmonic measure, it follows that

$$\int_J \phi(\zeta) \, d\omega_{F_\infty}(q(z), \zeta) = \int_J \phi(q(\zeta)) \, d\omega_{F_\infty}(z, \zeta) \quad (z \in F_\infty).$$

As this equation holds whenever ϕ is continuous, it remains true if ϕ is a bounded Borel function. Taking $\phi = 1_B$ gives the result. □

This can used to study ergodicity properties of q on the Julia set.

6.5. COMPLEX DYNAMICS

Corollary 6.5.7 *Let q, F_∞, J be as in the Theorem, and let ω be the Borel probability measure on J given by*

$$\omega(B) = \omega_{F_\infty}(\infty, B) \quad (B \in \mathcal{B}(J)).$$

Then:

(a) *ω is q-invariant: $\omega(q^{-1}(B)) = \omega(B)$ $(B \in \mathcal{B}(J))$;*

(b) *ω is q-ergodic: if $q^{-1}(B) = B$, then $\omega(B) = 0$ or 1;*

(c) *if $\phi: J \to \mathbb{R}$ is a bounded Borel function, then*

$$\lim_{n \to \infty} \frac{1}{n} \sum_{k=0}^{n-1} \phi(q^k(\zeta)) = \int_J \phi \, d\omega \quad \text{for } \omega\text{-a.e. } \zeta \in J.$$

Proof. (a) This follows by taking $z = \infty$ in Theorem 6.5.6.

(b) Suppose that B is a Borel subset of J such that $q^{-1}(B) = B$. Then applying Theorem 6.5.6 repeatedly gives

$$\omega_{F_\infty}(z, B) = \omega_{F_\infty}(q^n(z), B) \quad (z \in F_\infty),$$

and letting $n \to \infty$, it follows that

$$\omega_{F_\infty}(z, B) \equiv \omega(B) \quad (z \in F_\infty).$$

Now let C be a compact subset of B, and set

$$u(z) = \omega_{F_\infty}(z, C) \quad (z \in F_\infty).$$

Then u is harmonic on F_∞, and $u \leq \omega(B)$ there. Also, by Theorem 4.3.4 $\lim_{z \to \zeta} u(z) = 0$ for all $\zeta \in J \setminus C$. Hence by the two-constant theorem (Theorem 4.3.7),

$$u(z) \leq \omega(B)\omega_{F_\infty}(z, C) \quad (z \in F_\infty),$$

and in particular, putting $z = \infty$, we get $\omega(C) \leq \omega(B)\omega(C)$. Taking the supremum over all compact subsets C of B gives $\omega(B) \leq \omega(B)^2$, and hence $\omega(B) = 0$ or 1, as desired.

(c) This follows directly from (a) and (b) by applying the pointwise ergodic theorem ([65, Theorem 1.14]). □

Actually, by Theorem 4.3.14, the measure ω is none other than the equilibrium measure ν for J. Viewed in this light, it too has a purely dynamical interpretation.

Theorem 6.5.8 *Let q be a polynomial of degree $d \geq 2$, let J be its Julia set, and let $w \in J$. Let $(\mu_n)_{n \geq 1}$ be the sequence of Borel probability measures on J defined by*

$$\mu_n = \frac{1}{d^n} \sum_{q^n(\zeta_j) = w} \delta_{\zeta_j},$$

where δ_ζ denotes the unit mass at ζ, and the sum is taken over all d^n solutions ζ_j of the equation $q^n(\zeta) = w$, counted according to multiplicity. Then $\mu_n \xrightarrow{w^*} \nu$ as $n \to \infty$, where ν is the equilibrium measure for J.

Proof. Conjugating q by a polynomial $m(z) = \alpha z + \beta$ does not affect the truth of the result, so we may as well suppose that q is monic, since this will simplify the algebra. Note that by Theorem 6.5.1, this implies that $I(\nu) = \log c(J) = 0$.

We first show that the potentials of the μ_n satisfy

(6.26) $$\limsup_{n \to \infty} p_{\mu_n} \leq 0 \quad \text{on } J.$$

To see this, observe that since q is monic, we have

$$q^n(z) - w = \prod_{j=1}^{d^n} (z - \zeta_j),$$

where $\zeta_1, \ldots, \zeta_{d^n}$ are the solutions of $q^n(\zeta) = w$. Hence

$$p_{\mu_n}(z) = \frac{1}{d^n} \sum_{j=1}^{d^n} \log |z - \zeta_j| = \frac{1}{d^n} \log |q^n(z) - w| \quad (z \in \mathbb{C}).$$

This gives (6.26), since if $z \in J$ then $|q^n(z) - w|$ remains bounded as $n \to \infty$.

The next step is to show that if (μ_{n_j}) is any subsequence of (μ_n), then

(6.27) $$\limsup_{j \to \infty} p_{\mu_{n_j}} = 0 \quad \nu\text{-a.e. on } J.$$

By Frostman's theorem (Theorem 3.3.4) $p_\nu \geq I(\nu) = 0$ on $\mathbb{C}$, and hence, using Fatou's lemma and Fubini's theorem,

$$\int_J \limsup_{j \to \infty} p_{\mu_{n_j}} \, d\nu \geq \limsup_{j \to \infty} \int_J p_{\mu_{n_j}} \, d\nu = \limsup_{j \to \infty} \int_J p_\nu \, d\mu_{n_j} \geq 0.$$

Combining this with (6.26) we obtain (6.27).

The other fact we shall need is that $\operatorname{supp} \nu$ is the whole of J. To prove this, take $\zeta \in J$. Then by Theorem 4.2.4 and Corollary 6.5.5, $p_\nu(\zeta) = 0$. As

6.5. COMPLEX DYNAMICS

already remarked, $p_\nu \geq 0$ on $\mathbf{C}$, so this means that p_ν attains a minimum at ζ. Now if it were the case that $\zeta \notin \operatorname{supp}\nu$, then p_ν would be harmonic near ζ, and consequently, by the maximum principle, we would have $p_\nu \equiv 0$ on a neighbourhood of ζ. But every neighbourhood of ζ meets F_∞, where $p_\nu > 0$. We conclude therefore that $\zeta \in \operatorname{supp}\nu$.

Now we are ready to complete the proof. Suppose, for a contradiction, that (μ_n) is not weak*-convergent to ν. Then there exist a subsequence (μ_{n_j}), a continuous function $\phi\colon J \to \mathbf{R}$ and $\epsilon > 0$ such that

$$(6.28) \qquad \left| \int_J \phi \, d\mu_{n_j} - \int_J \phi \, d\nu \right| \geq \epsilon \quad \text{for all } j.$$

By Theorem A.4.2, replacing (μ_{n_j}) by a further subsequence, we can also suppose that $\mu_{n_j} \xrightarrow{w^*} \mu$, for some probability measure μ on J. Now by an argument similar to the proof of Lemma 3.3.3,

$$\limsup_{j \to \infty} p_{\mu_{n_j}}(z) \leq p_\mu(z) \quad (z \in \mathbf{C}).$$

Together with (6.27), this implies that $p_\mu \geq 0$ ν-a.e. on J. As p_μ is upper semicontinuous and $\operatorname{supp}\nu = J$, it follows that in fact $p_\mu \geq 0$ on J. The energy of μ therefore satisfies

$$I(\mu) = \int_J p_\mu \, d\mu \geq 0 = I(\nu),$$

so that μ is an equilibrium measure for J. Since the equilibrium measure is unique (Theorem 3.7.6), we conclude that $\mu = \nu$. But this then implies that $\mu_{n_j} \xrightarrow{w^*} \nu$, which contradicts (6.28), thereby proving the result. $\square$

Corollary 6.5.9 *If q, J, w are as in the Theorem, then*

$$J = \overline{\bigcup_{n \geq 0} q^{-n}(\{w\})}.$$

Proof. Since J is closed and completely invariant, we certainly have

$$\overline{\bigcup_{n \geq 0} q^{-n}(\{w\})} \subset J.$$

On the other hand, Theorem 6.5.8 implies that

$$\overline{\bigcup_{n \geq 0} q^{-n}(\{w\})} \supset \operatorname{supp}\nu,$$

and, as shown in the proof, $\operatorname{supp}\nu = J$, giving the reverse inclusion. $\square$

Potential theory also enters complex dynamics in quite another way. Frequently, instead of treating a single polynomial q in isolation, it is helpful to consider a whole family of polynomials $(q_\lambda)_{\lambda \in D}$, where D is a plane domain, and the family is holomorphic in the sense that the coefficients of q_λ are holomorphic in λ. It then turns out that several quantities associated to the Julia set J_λ of q_λ are subharmonic functions of λ. We shall illustrate this idea by applying it to Hausdorff dimension.

Let S be an arbitrary subset of $\mathbb{C}$. Given $\alpha > 0$, the α-*dimensional Hausdorff measure* of S is defined by

$$m_\alpha(S) = \lim_{\epsilon \to 0}\Big(\inf\Big\{\sum_j (\operatorname{diam} A_j)^\alpha : S \subset \cup_j A_j, \operatorname{diam} A_j < \epsilon\Big\}\Big),$$

the infimum being taken over all countable coverings of S by subsets A_j of $\mathbb{C}$ of diameter $< \epsilon$. This infimum obviously gets larger as ϵ decreases, so the limit always exists, though it may be infinite. In fact, there is always a number $\delta \in [0, 2]$ such that

$$m_\alpha(S) = \begin{cases} \infty, & \alpha < \delta, \\ 0, & \alpha > \delta. \end{cases}$$

This δ is called the *Hausdorff dimension* of S, denoted by $\dim(S)$. A nonempty open subset of $\mathbb{C}$ always has dimension 2, while a countable set, and more generally a Borel polar set, has dimension 0. In between, every (nonconstant) Lipschitz curve has dimension 1. On the other hand the Cantor set has a fractional dimension, namely $\log 2/\log 3$. For more about such 'fractals', and about Hausdorff dimension in general, see [29].

Julia sets are often quoted as examples of fractals, though their Hausdorff dimension is rather hard to calculate. In [31] Garber proved that for any polynomial q of degree $d \geq 2$,

$$\dim(J) \geq \frac{\log d}{\log(\sup_J |q'|)}, \tag{6.29}$$

and hence, in particular, the dimension of J is always strictly positive. At the other extreme, Shishikura [60] has shown that there exist quadratic polynomials q whose Julia sets have dimension equal to 2 (this is still slightly weaker than saying that J has positive area).

There is one case where we have an exact formula for $\dim(J)$, namely when q is *hyperbolic*. This means that all its critical points (points where $q'(z) = 0$) and their iterates under q stay a positive distance from the Julia set. One can show that this is equivalent to the existence of constants $C > 0$ and $\gamma > 1$ such that

$$\inf_J |(q^n)'| \geq C\gamma^n \quad (n \geq 1). \tag{6.30}$$

6.5. COMPLEX DYNAMICS

In this case, the Hausdorff dimension of the Julia set is given by the *Bowen–Ruelle–Manning formula*:

$$(6.31) \qquad \dim(J) = \sup_{\mu} \left(\frac{e_\mu(q)}{\int_J \log |q'| \, d\mu} \right),$$

where the supremum is taken over all Borel probability measures μ on J which are q-invariant, and $e_\mu(q)$ denotes the entropy of q with respect to μ (more about this in a moment). To explain this formula in detail would take us too far afield, but briefly the idea is the following. To compute Hausdorff dimension one generally has to exploit the self-similarity properties of a set. Now although a Julia set is not usually self-similar, one can use (6.30) to show that if q is hyperbolic then the restriction $q \colon J \to J$ is an approximate self-similarity, which, together with some ergodic theory, eventually leads to the formula (6.31). For more details see [52] and [45]. The latter also explains the concept of entropy. For us it will be enough to know that $0 \leq e_\mu(q) \leq \log d$, where d is the degree of q, and that $e_\nu(q) = \log d$ if and only if ν is the equilibrium measure for J, which we saw was q-invariant in Corollary 6.5.7. Thus (6.31) actually implies (6.29) when q is hyperbolic.

Unfortunately, in many cases the Bowen–Ruelle–Manning formula is too complicated to apply directly. It is then that it becomes useful to consider holomorphic families, as exemplified by the following theorem.

Theorem 6.5.10 *Let $(q_\lambda)_{\lambda \in D}$ be a holomorphic family of polynomials of constant degree $d \geq 2$, and assume further that each q_λ is hyperbolic, and that D is simply connected. Then the Hausdorff dimension of the Julia set J_λ of q_λ is given by*

$$(6.32) \qquad \frac{1}{\dim(J_\lambda)} = \inf_{h \in \mathcal{H}} h(\lambda) \quad (\lambda \in D),$$

where $\mathcal{H}$ is a family of harmonic functions on D with the following properties:

(a) *every $h \in \mathcal{H}$ satisfies $h \geq \frac{1}{2}$ on D;*

(b) *for each $h \in \mathcal{H}$ there exists a constant $C \geq 1$ such that*

$$(6.33) \qquad C \inf_{J_\lambda} \left(\frac{\log |q'_\lambda|}{\log d} \right) \leq h(\lambda) \leq C \sup_{J_\lambda} \left(\frac{\log |q'_\lambda|}{\log d} \right) \quad (\lambda \in D).$$

Moreover, there exists $h_0 \in \mathcal{H}$ such that:

(c) *$h_0 \geq 1$ on D, with equality at λ if and only if no finite critical point of q_λ is attracted to ∞;*

(d) *h_0 satisfies (6.33) with $C = 1$.*

Before proving this theorem, let us read off some simple corollaries.

Corollary 6.5.11 *Under the assumptions of the Theorem, we have*

$$(6.34) \qquad \frac{1/\dim(J_{\lambda_1}) - \frac{1}{2}}{1/\dim(J_{\lambda_2}) - \frac{1}{2}} \leq \tau_D(\lambda_1, \lambda_2) \qquad (\lambda_1, \lambda_2 \in D),$$

where τ_D denotes the Harnack distance on D. Consequently, the function $\lambda \mapsto \dim(J_\lambda)$ is continuous on D.

Proof. Given $h \in \mathcal{H}$, the function $h - \frac{1}{2}$ is positive and harmonic on D, so by the definition of Harnack distance (Definition 1.3.4)

$$h(\lambda_1) - \tfrac{1}{2} \leq \tau_D(\lambda_1, \lambda_2)(h(\lambda_2) - \tfrac{1}{2}) \qquad (\lambda_1, \lambda_2 \in D).$$

Taking the infimum over all $h \in \mathcal{H}$ yields the inequality (6.34). This implies that $\dim(J_\lambda)$ is continuous, because $\log \tau_D$ is a continuous semimetric on D (see Theorem 1.3.8). □

We remark in passing that, still under the assumptions of the Theorem, one can show that $\dim(J_\lambda)$ is actually real-analytic in λ—see [59] for more details. The next corollary gives another property of this function.

Corollary 6.5.12 *Under the assumptions of the Theorem, the function $\lambda \mapsto 1/\dim(J_\lambda)$ is superharmonic on D, and $\lambda \mapsto \dim(J_\lambda)$ is subharmonic on D.*

Proof. As $1/\dim(J_\lambda)$ is an infimum of harmonic functions, it must satisfy the supermean property, and we have just seen that it is continuous, hence it is superhamonic on D. Also, $\dim(J_\lambda) = \psi(-1/\dim(J_\lambda))$, where $\psi(x) = -1/x$, which is convex and increasing on $x < 0$. Hence by Theorem 2.6.3 $\dim(J_\lambda)$ is subharmonic on D. □

Proof of Theorem 6.5.10 (sketch). Applying the Bowen–Ruelle–Manning formula, we have

$$(6.35) \qquad \dim(J_\lambda) = \sup_{\mu \in \mathcal{M}_\lambda} \left(\frac{e_\mu(q_\lambda)}{\int_{J_\lambda} \log |q'_\lambda| \, d\mu} \right) \qquad (\lambda \in D),$$

where $\mathcal{M}_\lambda$ denotes the set of all q_λ-invariant Borel probability measures on J_λ.

We need to know how the various quantities on the right-hand side vary with λ, and for this it is necessary to invoke another consequence of hyperbolicity, due to Mañé, Sad and Sullivan [43]. Fix $\lambda_0 \in D$. They showed that, under the conditions of Theorem 6.5.10, there is a family of maps $(\phi_\lambda)_{\lambda \in D}$ such that:

6.5. COMPLEX DYNAMICS

(i) ϕ_λ is a homeomorphism of J_{λ_0} onto J_λ, for each $\lambda \in D$;

(ii) $\phi_{\lambda_0} = $ identity on J_{λ_0};

(iii) $\lambda \mapsto \phi_\lambda(z)$ is holomorphic on D, for each $z \in J_{\lambda_0}$;

(iv) $q_\lambda = \phi_\lambda \circ q_{\lambda_0} \circ \phi_\lambda^{-1}$ on J_λ, for each $\lambda \in D$.

Properties (i)–(iii) are summarized by saying that $(\phi_\lambda)_{\lambda \in D}$ is a *holomorphic motion*. Property (iv) tells us that the action of q_λ on J_λ is conjugate to the action of q_{λ_0} on J_{λ_0}.

This last property implies that, for each $\lambda \in D$, the map $\mu \mapsto \mu\phi_\lambda^{-1}$ is a bijection between $\mathcal{M}_{\lambda_0}$ and $\mathcal{M}_\lambda$. Moreover this map preserves entropy, because entropy is conjugation-invariant. Hence (6.35) now becomes

$$\dim(J_\lambda) = \sup_{\mu \in \mathcal{M}_{\lambda_0}} \left(\frac{e_\mu(q_{\lambda_0})}{\int_{J_{\lambda_0}} \log |q'_\lambda \circ \phi_\lambda| \, d\mu} \right) \quad (\lambda \in D).$$

Thus if we define $\mathcal{H}$ to be the collection of all functions $h \colon D \to \mathbb{R}$ of the form

$$(6.36) \qquad h(\lambda) = \frac{\int_{J_{\lambda_0}} \log |q'_\lambda \circ \phi_\lambda| \, d\mu}{e_\mu(q_{\lambda_0})} \quad (\lambda \in D),$$

where $\mu \in \mathcal{M}_{\lambda_0}$ and $e_\mu(q_{\lambda_0}) > 0$, then each $h \in \mathcal{H}$ is harmonic on D, and (6.32) holds.

It remains to show that $\mathcal{H}$ has properties (a)–(d). Property (a) is clear, because each $h \in \mathcal{H}$ satisfies $h(\lambda) \geq 1/\dim(J_\lambda)$, and certainly $\dim(J_\lambda) \leq 2$. For property (b), we go back to the formula (6.36). Let $\mu \in \mathcal{M}_{\lambda_0}$ and set $C = (\log d)/e_\mu(q_{\lambda_0})$, so that by our earlier remarks about entropy $C \geq 1$. Then the corresponding function $h \in \mathcal{H}$ is given by

$$h(\lambda) = \frac{C}{\log d} \int_{J_{\lambda_0}} \log |q'_\lambda \circ \phi_\lambda| \, d\mu = \frac{C}{\log d} \int_{J_\lambda} \log |q'_\lambda| \, d(\mu\phi_\lambda^{-1}) \quad (\lambda \in D),$$

and the inequality (6.33) follows from this because $\mu\phi_\lambda^{-1}$ is a probability measure on J_λ.

For properties (c) and (d), we take h_0 to be the function in $\mathcal{H}$ corresponding to the equilibrium measure ν for J_{λ_0}. Since $e_\nu(q_{\lambda_0}) = \log d$, the working just above shows that h_0 satisfies (6.33) with $C = 1$, which gives property (d). To prove that (c) holds, we need to express h_0 in yet another way. Fix $\lambda \in D$, and write $q_\lambda(z) = \sum_{j=0}^{d} a_j z^j$. Then we have $q'_\lambda(z) = d a_d (z - c_1) \cdots (z - c_{d-1})$, where $c_1, \ldots, c_{d-1}$ are the (finite) critical points of q_λ, and hence

$$\begin{aligned}
h_0(\lambda) &= \frac{1}{\log d} \int_{J_\lambda} \log |q'_\lambda|\, d(\nu \phi_\lambda^{-1}) \\
&= \frac{1}{\log d} \left(\log d + \log |a_d| + \sum_{k=1}^{d-1} p_{\nu \phi_\lambda^{-1}}(c_k) \right) \\
&= 1 + \frac{1}{\log d} \sum_{k=1}^{d-1} (p_{\nu \phi_\lambda^{-1}}(c_k) - \log c(J_\lambda)),
\end{aligned}$$

the last line coming from Theorem 6.5.1. Now $e_{\nu \phi_\lambda^{-1}}(q_\lambda) = e_\nu(q_{\lambda_0}) = \log d$, so $\nu \phi_\lambda^{-1}$ is the equilibrium measure for J_λ. By Frostman's theorem it follows that for each c_k we have $p_{\nu \phi_\lambda^{-1}}(c_k) \geq \log c(J_\lambda)$, with equality unless c_k lies in the attracting basin of ∞ of q_λ. This implies property (c). □

To illustrate the use of Theorem 6.5.10, here is a simple application.

Theorem 6.5.13 *Let $q_\lambda(z) = z^2 + \lambda z$, and let J_λ be its Julia set. Then*

$$1 \leq \dim(J_\lambda) \leq 1 + |\lambda| \qquad (|\lambda| < 1).$$

Proof. Observe that $q_\lambda(0) = 0$ and $q'_\lambda(0) = |\lambda|$, so that if $|\lambda| < 1$ then 0 is an attracting fixed point of q_λ. The general theory then tells us that $0 \notin J_\lambda$ and that 0 attracts a critical point of q_λ (see [13, Theorem 9.7.5]). Since q_λ has only one (finite) critical point, it must be attracted to 0, and hence stays away from the Julia set. Thus q_λ is hyperbolic for $|\lambda| < 1$, and Theorem 6.5.10 is applicable.

For the lower bound, observe that the function h_0 from part (c) of Theorem 6.5.10 satisfies $h_0 \equiv 1$, and hence

$$\frac{1}{\dim(J_\lambda)} = \inf_{h \in \mathcal{H}} h(\lambda) \leq h_0(\lambda) = 1 \qquad (|\lambda| < 1).$$

For the upper bound, applying Corollary 6.5.11 with $\lambda_1 = 0$ and $\lambda_2 = \lambda$ gives

$$\frac{1/\dim(J_0) - \frac{1}{2}}{1/\dim(J_\lambda) - \frac{1}{2}} \leq \tau_{\Delta(0,1)}(0, \lambda) = \frac{1 + |\lambda|}{1 - |\lambda|}.$$

Since $q_0(z) = z^2$, it follows that $\dim(J_0) = 1$, and so we deduce that

$$\frac{1}{\dim(J_\lambda)} \geq \frac{1}{2} \frac{1 - |\lambda|}{1 + |\lambda|} + \frac{1}{2} = \frac{1}{1 + |\lambda|} \qquad (|\lambda| < 1),$$

which gives the result. □

6.5. COMPLEX DYNAMICS

Every quadratic polynomial is conjugate (via a suitable $m(z) = \alpha z + \beta$) to a unique polynomial of the form $q_c(z) = z^2 + c$, where $c \in \mathbb{C}$. Thus one can regard the c-plane as parametrizing the conjugacy classes of quadratic polynomials. The dynamics of q_c is greatly influenced by what happens to its critical point, 0, and it is particularly important whether this point lies in the attracting basin of ∞. Accordingly, we define

$$M = \{c \in \mathbb{C} : q_c^n(0) \not\to \infty\}.$$

This is the celebrated *Mandelbrot set*. One can show that it is a connected, simply connected, compact set of capacity 1, though in many other ways its structure is very rich and complex. For more information, see for example [23, Chapter VIII].

A simple calculation shows that $z^2 + \lambda z$ is conjugate to $z^2 + c$, where $c = \lambda/2 - \lambda^2/4$. Thus the polynomials in Theorem 6.5.13 correspond to the c-values in the cardioid

$$W = \{\lambda/2 - \lambda^2/4 : |\lambda| < 1\}.$$

(One can show that in fact W is a component of the interior of M, the so-called *main cardioid* of the Mandelbrot set.) Thus from Theorem 6.5.13 we immediately obtain the following corollary.

Corollary 6.5.14 *Let $q_c(z) = z^2 + c$, and let J_c be its Julia set. Then*

$$1 \leq \dim(J_c) \leq 1 + |1 - \sqrt{1 - 4c}| \qquad (c \in W). \quad \square$$

The lower bound in this result is easily explained by the fact that J_c is connected for all $c \in W$. More generally, one can show that J_c is connected for all $c \in M$, while on the other hand, J_c is totally disconnected for all $c \in \mathbb{C} \setminus M$. Our last application concerns this complementary region.

Theorem 6.5.15 *Let $q_c(z) = z^2 + c$, and let J_c be its Julia set. Then*

$$\left(\frac{g(c)}{2\log 2} + 1\right)^{-1} \leq \dim(J_c) \leq \left(\frac{g(c)}{2\log 2} + \frac{1}{e^{-g(c)/2} + 1}\right)^{-1} \qquad (c \in \mathbb{C} \setminus M),$$

where $g(c) = g_{\mathbb{C}_\infty \setminus M}(c, \infty)$.

One can show that

$$g(c) = \lim_{n \to \infty} 2^{-n} \log |q_c^n(c)| \quad (c \in \mathbb{C} \setminus M),$$

the convergence being quite rapid (see [62, §158]), so the bounds in Theorem 6.5.15 are readily computable.

Proof of Theorem 6.5.15 (sketch). If $c \in \mathbb{C} \setminus M$, then $q_c^n(0) \to \infty$ as $n \to \infty$, so this time it is clear that q_c is hyperbolic. Unfortunately, however, we cannot apply Theorem 6.5.10 directly, because the domain $\mathbb{C} \setminus M$ is not simply connected. Simple connectedness was needed in the proof of Theorem 6.5.10 in order for the result of Mañé, Sad and Sullivan to be applicable, and without it the argument breaks down. Indeed, there is no holomorphic function ψ on $\mathbb{C} \setminus M$ such that $\psi(c) \in J_c$ for all $c \in \mathbb{C} \setminus M$. We can demonstrate this as follows. If $|z| > |c|^{1/2} + 1$ or $|z| < |c|^{1/2} - 1$, then elementary estimates show that $q_c^n(z) \to \infty$ as $n \to \infty$, and hence

$$(6.37) \qquad J_c \subset \{z : |c|^{1/2} - 1 \leq |z| \leq |c|^{1/2} + 1\} \quad (c \in \mathbb{C}).$$

Consequently, any function ψ for which $\psi(c) \in J_c$ for all $c \in \mathbb{C} \setminus M$ must satisfy $\psi(c) = |c|^{1/2} + O(1)$ as $c \to \infty$, and this is impossible if ψ is holomorphic.

To overcome this problem, we reparametrize the family $(q_c)_{c \in \mathbb{C} \setminus M}$. Set $D = \{\lambda \in \mathbb{C} : |\lambda| > 1\}$. Since $\mathbb{C}_\infty \setminus M$ is simply connected, there exists a conformal mapping $f : D \to \mathbb{C} \setminus M$ such that $\lim_{\lambda \to \infty} f(\lambda) = \infty$. Moreover, as M has capacity 1, the mapping f can be chosen so that

$$(6.38) \qquad f(\lambda) = \lambda + O(1) \quad \text{as } \lambda \to \infty.$$

For $\lambda \in D$, we define

$$\widetilde{q}_\lambda(z) = q_{f(\lambda^2)}(z) = z^2 + f(\lambda^2),$$

so that the corresponding Julia set is given by

$$\widetilde{J}_\lambda = J_{f(\lambda^2)}.$$

With this parametrization, one can find a holomorphic function ψ_0 on D such that $\psi_0(\lambda)$ is a repelling fixed point of $\widetilde{q}_\lambda$ for each $\lambda \in D$, and then, for each $n \geq 1$, distinct holomorphic functions $(\psi_{n,j})_{j=1}^{2^n}$ on D with $\widetilde{q}_\lambda^n(\psi_{n,j}(\lambda)) = \psi_0(\lambda)$ $(\lambda \in D)$. (The square roots that arise are all single-valued because $f(\lambda^2) = \lambda^2 + O(|\lambda|)$ as $\lambda \to \infty$.) Proceeding as in [43], one can now construct a holomorphic motion $(\phi_\lambda)_{\lambda \in D}$ satisfying the properties (i)–(iv) in the proof of Theorem 6.5.10, but with q_λ, J_λ replaced by $\widetilde{q}_\lambda, \widetilde{J}_\lambda$ throughout. The rest of the proof then goes through as before, and so Theorem 6.5.10 is applicable to the family $(\widetilde{q}_\lambda)_{\lambda \in D}$.

For the lower bound, we consider once again the function $h_0 \in \mathcal{H}$ from parts (c) and (d) of Theorem 6.5.10. From (c) we have $h_0 \geq 1$ on D, and combining (d) with (6.37) and (6.38) gives

$$h_0(\lambda) = \frac{\log(2|f(\lambda^2)|^{1/2})}{\log 2} + o(1) = \frac{\log |\lambda|}{\log 2} + 1 + o(1) \quad \text{as } \lambda \to \infty.$$

6.5. COMPLEX DYNAMICS

Thus if we set
$$k_0(\lambda) = h_0(\lambda) - \frac{\log|\lambda|}{\log 2} \quad (\lambda \in D),$$
then k_0 is harmonic on D, with $\lim_{\lambda \to \infty} k_0(\lambda) = 1$ and $\liminf_{\lambda \to \zeta} k_0(\lambda) \geq 1$ for each $\zeta \in \partial D \setminus \{\infty\}$. Removing the singularity at ∞, and applying the maximum principle, it follows that $k_0 \equiv 1$ on D, and hence
$$h_0(\lambda) = \frac{\log|\lambda|}{\log 2} + 1 \quad (\lambda \in D).$$
Now with $c = f(\lambda^2)$ we have
$$(6.39) \qquad \log|\lambda| = (\log|f^{-1}(c)|)/2 = g(c)/2,$$
where $g(c) = g_{\mathbb{C}_\infty \setminus M}(c, \infty)$. Therefore
$$\frac{1}{\dim(J_c)} = \frac{1}{\dim(\widetilde{J}_\lambda)} \leq h_0(\lambda) = \frac{g(c)}{2\log 2} + 1 \quad (c \in \mathbb{C} \setminus M).$$
This proves the lower bound.

For the upper bound, we repeat the above analysis with a general $h \in \mathcal{H}$. This time parts (a) and (b) of Theorem 6.5.10 tell us that $h \geq \frac{1}{2}$ on D and
$$h(\lambda) = C\left(\frac{\log|\lambda|}{\log 2} + 1\right) + o(1) \quad \text{as } \lambda \to \infty,$$
where C is a constant with $C \geq 1$. Thus if we set
$$k(\lambda) = h(\lambda) - C\frac{\log|\lambda|}{\log 2} \quad (\lambda \in D),$$
then k is harmonic on D, with $\lim_{\lambda \to \infty} k(\lambda) = C$ and $\liminf_{\lambda \to \zeta} k(\lambda) \geq \frac{1}{2}$ for each $\zeta \in \partial D \setminus \{\infty\}$. Removing the singularity at ∞, and applying Harnack's inequality to $(k - \frac{1}{2})$, we deduce that
$$k(\lambda) - \tfrac{1}{2} \geq (C - \tfrac{1}{2})\frac{|\lambda| - 1}{|\lambda| + 1} \quad (\lambda \in D),$$
and hence, since $C \geq 1$,
$$h(\lambda) \geq \frac{\log|\lambda|}{\log 2} + \frac{1}{1/|\lambda| + 1} \quad (\lambda \in D).$$
With $c = f(\lambda^2)$, we use (6.39) once more, and obtain
$$\frac{1}{\dim(J_c)} = \frac{1}{\dim(\widetilde{J}_\lambda)} = \inf_{h \in \mathcal{H}} h(\lambda) \geq \frac{g(c)}{2\log 2} + \frac{1}{e^{-g(c)/2} + 1} \quad (c \in \mathbb{C} \setminus M).$$
This gives the upper bound, and completes the proof. $\square$

Notes on Chapter 6

§6.1

Thorin's complex-variable proof of Riesz's theorem has been generalized by Calderón to provide a method for interpolating between general Banach spaces. For more about this see the book of Bergh and Löfstrom [15], which also contains an account of the history of the theorem.

§6.2

Looking back at the proof of Theorem 6.2.1, the only property of p used, other than homogeneity, is the fact that $\log |p(f_1(\lambda), \ldots, f_n(\lambda))|$ is subharmonic whenever $f_1, \ldots, f_n$ are holomorphic functions, i.e. that $\log |p|$ is *plurisubharmonic* on $\mathbb{C}^n$. Thus the result extends to homogeneous plurisubharmonic functions, and in fact it can be generalized (appropriately) even to inhomogeneous ones of finite order. For more about this, as well as a partial converse to Theorem 6.2.1, see [55].

Corollary 6.2.4 was proved for connected sets by Pommerenke in [50], using methods from the theory of univalent functions. The same paper also considers *lower* bounds for capacity in terms of the n-th diameter.

Corollary 6.2.5 was suggested to the author by Bernard Aupetit. The inequality for Minkowski sums was again proved for connected sets by Pommerenke [49], who used it to give a new proof of the theorem of Pólya and Schiffer that, if K is a compact convex set of perimeter P, then $c(K) \geq P/8$.

§6.3

The Bernstein–Walsh theorem (Theorem 6.3.1) has many proofs. A recent survey of Bernstein-type theorems appears in [9].

Keldysh's theorem (Theorem 6.3.3) too can be generalized in several directions. For more about this we refer to [8].

§6.4

The proof given here of Johnson's theorem (Theorem 6.4.4) is from [54]. The books of Aupetit [5, 6] contain several other applications of Vesentini's theorem and its generalizations, as well as much useful background information on spectral theory.

One can interpret Rudin's theorem (Theorem 6.4.8) as saying that the set $\Phi_A \setminus \partial_A$ carries an analytic structure with respect to which the Gelfand transforms of elements of A are holomorphic. For a while it was thought that the same might be true of an arbitrary commutative Banach algebra with identity. Even though this turns out not to be true in general, there are some partial results in this direction, for which Wermer's theorem and its generalizations are again the tools. A fuller account of this may be found in [67, Chapter 11] and [5, Chapter 3, §5].

Theorem 6.4.9 is from [61, Chapter 5], although our proof uses the Phragmén–Lindelöf theorem instead of the more difficult Ahlfors–Heins theorem. The same chapter also contains two further results of Esterle about the growth of holomorphic semigroups in radical Banach algebras.

§6.5

Theorem 6.5.8 is due to Brolin [19], who showed that in fact the same conclusion holds for every $w \in \mathbb{C}$ with at most one exception, namely when the singleton $\{w\}$ is completely invariant. This result suggests a way of drawing pictures of J on a computer: pick a point w, and 'iterate backwards'. However, quite apart from the problem that the number of points in $q^{-n}(\{w\})$ grows exponentially with n, Theorem 6.5.8 itself hints at another possible source of difficulty. Parts of J where ν is too thinly spread out may show up too rarely amongst the $q^{-n}(\{w\})$ to be visible. To overcome this the algorithm has to be modified to try to take into account the distribution of ν. For more details see [47, Chapter 2].

Theorem 6.5.10 and its applications are from [56]. A similar technique was recently used by Astala [4] to solve a long-standing problem about area distortion of quasiconformal mappings.

Finally ...

It is perhaps worth mentioning that several of the results in this chapter are closely related to the theory of *analytic multifunctions*. This is a class of set-valued functions of a complex variable which displays many properties reminiscent of holomorphic and subharmonic functions, among them: a maximum principle, a Liouville theorem, an open mapping theorem, an identity principle, an argument principle, a Rouché theorem, a two-constant theorem, a Brelot–Cartan theorem, and several Picard-type theorems. As an example, if f is a holomorphic function from an open subset of $\mathbb{C}$ into a Banach algebra A, then its spectrum $\sigma(f)$ is an analytic multifunction— a generalization of Vesentini's theorem (Theorem 6.4.2) which has many applications. Other examples arise in connection with interpolation (§6.1), capacity (§6.2), analytic structure in the character space (§6.4) and Julia sets of holomorphic families (§6.5), as well as in several complex variables and martingale theory. A brief introduction to analytic multifunctions can be found in the final chapter of [6].

Appendix A

Borel Measures

By the term *Borel measure* is meant a positive measure on the Borel σ-algebra of a topological space. In this appendix we provide proofs of the various results about Borel measures that are used in the course of the book.

A.1 Supports

Definition A.1.1 Let μ be a Borel measure on a topological space X. The *support* of μ, denoted $\operatorname{supp}\mu$, is the set of $x \in X$ such that $\mu(U) > 0$ for each open neighbourhood U of x.

Theorem A.1.2 *Assume that X has a countable base of open sets. Then $\operatorname{supp}\mu$ is the smallest closed subset F of X such that $\mu(X \setminus F) = 0$.*

Proof. Let $(U_n)_{n \geq 1}$ be a countable base of open sets for X. Then

$$X \setminus \operatorname{supp}\mu = \bigcup_n \{U_n : \mu(U_n) = 0\},$$

so $\operatorname{supp}\mu$ is closed in X, and $\mu(X \setminus \operatorname{supp}\mu) = 0$. Also, if F is a closed subset of X with $\mu(X \setminus F) = 0$, then necessarily

$$X \setminus F \subset \bigcup_n \{U_n : \mu(U_n) = 0\},$$

and so $\operatorname{supp}\mu \subset F$. $\square$

A.2 Regularity

Definition A.2.1 Let μ be a Borel measure on a topological space X. Then μ is called *regular* if each Borel set B has the following property: given $\epsilon > 0$, there exist an open set U and a closed set F such that $F \subset B \subset U$ and $\mu(U \setminus F) < \epsilon$.

Under appropriate conditions, regularity is automatic.

Theorem A.2.2 *If μ is a finite Borel measure on a metric space X, then μ is regular.*

Proof. Denote by $\mathcal{A}$ the class of Borel sets which have the stated regularity property. Evidently $\varnothing, X \in \mathcal{A}$, and also $\mathcal{A}$ is closed under taking complements. We next show that it is closed under countable unions. Suppose that $(B_n)_{n \geq 1} \in \mathcal{A}$, and let $\epsilon > 0$. Then for each $n \geq 1$ there exist an open set U_n and a closed set F_n such that $F_n \subset B_n \subset U_n$ and $\mu(U_n \setminus F_n) < \epsilon/2^{n+1}$. Put $U = \cup_1^\infty U_n$ and $F = \cup_1^N F_n$, where N is chosen large enough so that $\mu(\cup_1^\infty F_n \setminus F) < \epsilon/2$ (this is possible since $\mu(X) < \infty$). Then U is open, F is closed, $F \subset \cup_1^\infty B_n \subset U$ and

$$\mu(U \setminus F) \leq \sum_1^\infty \mu(U_n \setminus F_n) + \mu\left(\bigcup_1^\infty F_n \setminus F\right) < \epsilon/2 + \epsilon/2 = \epsilon,$$

so $\cup_1^\infty B_n \in \mathcal{A}$. Thus $\mathcal{A}$ is a σ-algebra. Moreover $\mathcal{A}$ contains the open sets, since each such set is the union of an increasing sequence of closed sets. Hence $\mathcal{A}$ contains all the Borel sets, which proves the result. $\square$

A consequence of regularity is the following approximation theorem.

Theorem A.2.3 (Vitali–Carathéodory Theorem) *Suppose that μ is a regular Borel measure on a topological space X, and that $\phi \colon X \to \mathbb{R}$ is an integrable function. Then, given $\epsilon > 0$, there exist an upper semicontinuous function $\psi_u \colon X \to [-\infty, \infty)$ and a lower semicontinuous function $\psi_l \colon X \to (-\infty, \infty]$ such that $\psi_u \leq \phi \leq \psi_l$ and $\int_X (\psi_u - \psi_l)\, d\mu < \epsilon$.*

Proof. It suffices to consider the case $\phi \geq 0$ (the general case then follows by writing $\phi = \phi^+ - \phi^-$ and treating ϕ^+ and ϕ^- separately). Since ϕ is the limit of an increasing sequence of simple functions s_n, it is also the sum of the non-negative simple functions $t_n := s_n - s_{n-1}$. Hence it can be written in the form

$$\phi = \sum_1^\infty c_n 1_{B_n},$$

where $c_n \geq 0$, and 1_{B_n} denotes the characteristic function of the Borel set B_n. By regularity, for each $n \geq 1$ there exist an open set U_n and a closed set F_n with $F_n \subset B_n \subset U_n$ and $c_n \mu(U_n \setminus F_n) < \epsilon/2^{n+1}$. Also, since

$$\sum_1^\infty c_n \mu(B_n) = \int_X \phi \, d\mu < \infty,$$

there exists an integer $N \geq 1$ such that $\sum_{N+1}^\infty c_n \mu(B_n) < \epsilon/2$. Set

$$\psi_u = \sum_1^N c_n 1_{F_n} \quad \text{and} \quad \psi_l = \sum_1^\infty c_n 1_{U_n}.$$

Then ψ_u is upper semicontinuous, ψ_l is lower semicontinuous, $\psi_u \leq \phi \leq \psi_l$ and

$$\begin{aligned}
\int_X (\psi_u - \psi_l) \, d\mu &\leq \sum_1^N c_n \mu(U_n \setminus F_n) + \sum_{N+1}^\infty c_n \mu(U_n) \\
&\leq \sum_1^\infty c_n \mu(U_n \setminus F_n) + \sum_{N+1}^\infty c_n \mu(B_n) \\
&< \epsilon/2 + \epsilon/2 = \epsilon.
\end{aligned}$$

This completes the proof. □

A.3 Radon Measures

Definition A.3.1 A Borel measure μ on a topological space X is called a *Radon measure* if $\mu(K) < \infty$ for each compact subset K of X.

Given a topological space X, we denote by $C_c(X)$ the vector space of continuous functions $\phi \colon X \to \mathbb{R}$ which have compact support. Each Radon measure μ on X gives rise to a linear functional Λ on $C_c(X)$ via

(A.1) $$\Lambda(\phi) = \int_X \phi \, d\mu \quad (\phi \in C_c(X)).$$

This linear functional is positive in the sense that $\Lambda(\phi) \geq 0$ whenever $\phi \geq 0$. For certain spaces X there is an important converse.

Theorem A.3.2 (Riesz Representation Theorem) *Let X be a metric space possessing a compact exhaustion. If Λ is a positive linear functional on $C_c(X)$, then there exists a unique Radon measure μ on X such that (A.1) holds.*

To say that X has a compact exhaustion means that there exists a sequence of compact subsets $(K_n)_{n \geq 1}$ such that $K_n \subset \text{int}(K_{n+1})$ for each n and $\cup_n K_n = X$. For example, open subsets of $\mathbb{C}$ have this property, as (trivially) do all compact metric spaces. Note that every metric space with a compact exhaustion is automatically locally compact.

The proof of the Riesz representation theorem splits into two parts: existence and uniqueness. We begin with uniqueness, for which we need the following regularity lemma.

Lemma A.3.3 *Let X be as in the Theorem, and let μ be a Radon measure on X. Then for each Borel subset B of X,*

$$\mu(B) = \sup\{\mu(K) : \text{ compact } K \subset B\}.$$

Proof. Let $\alpha < \mu(B)$. If $(K_n)_{n \geq 1}$ is a compact exhaustion of X, then $\mu(B \cap K_n)$ increases to $\mu(B)$, so there exists $N \geq 1$ such that $\mu(B \cap K_N) > \alpha$. Applying Theorem A.2.2 to the finite measure $\mu|_{K_N}$, we deduce that there exists a compact subset K of $B \cap K_N$ such that $\mu(K) > \alpha$. As this is true for each $\alpha < \mu(B)$, the result follows. $\square$

Proof of Theorem A.3.2 (Uniqueness). Let μ_1 and μ_2 be two Radon measures on X which satisfy (A.1). Then we have

$$\int_X \phi \, d\mu_1 = \int_X \phi \, d\mu_2 \quad (\phi \in C_c(X)).$$

If K is a compact subset of X, then taking ϕ to be of the form

$$\phi(x) := \max(0, 1 - n \, \text{dist}(x, K)) \quad (x \in X)$$

and letting $n \to \infty$, we deduce that $\mu_1(K) = \mu_2(K)$. From Lemma A.3.3 it follows that $\mu_1(B) = \mu_2(B)$ for each Borel set B, i.e. $\mu_1 = \mu_2$. This completes the proof of uniqueness. $\square$

We now turn to the proof of the existence of μ, which is rather longer. It is convenient first to introduce some notation. Given a compact subset K of X, we write $K \prec \phi$ to mean that

$$\phi \in C_c(X), \quad 0 \leq \phi \leq 1, \quad \phi = 1 \text{ on } K.$$

Also, given an open subset U of X, we write $\phi \prec U$ to mean that

$$\phi \in C_c(X), \quad 0 \leq \phi \leq 1, \quad \text{supp} \, \phi \subset U.$$

We shall need the following lemma on partitions of unity.

A.3. RADON MEASURES

Lemma A.3.4 *Let X be as in the Theorem, let K be a compact subset of X, and let $U_1, \ldots, U_N$ be an open cover of K. Then there exist $\psi_1, \ldots, \psi_N \in C_c(X)$ such that $\psi_n \prec U_n$ for each n and $K \prec \sum_1^N \psi_n$.*

Proof. Each $x \in K$ has an open neighbourhood whose closure is a compact subset of one of the U_n. Finitely many such neighbourhoods cover K. Denoting by V_n the union of those whose closures lie inside U_n, we obtain an open cover $V_1, \ldots, V_N$ of K such that $\overline{V}_n$ is a compact subset of U_n for every n. Define

$$\psi_n(x) = \frac{\operatorname{dist}(x, X \setminus V_n)}{\operatorname{dist}(x, K) + \sum_1^N \operatorname{dist}(x, X \setminus V_k)} \quad (x \in X).$$

Then $\psi_n \prec U_n$ for each n and $K \prec \sum_1^N \psi_n$. □

Proof of Theorem A.3.2 (Existence). Define a set-function μ^* on X as follows: if U is open in X, then

$$\mu^*(U) := \sup\{\Lambda(\phi) : \phi \prec U\},$$

and if E is an arbitrary subset of X, then

$$\mu^*(E) := \inf\{\mu^*(U) : \text{ open } U \supset E\}.$$

Although this apparently gives two definitions for $\mu^*(U)$, it is easily seen that they are consistent.

The first step is to show that μ^* is an *outer measure*, i.e. that

(a) $\mu^*(\emptyset) = 0$,

(b) $\mu^*(E_1) \leq \mu^*(E_2)$ whenever $E_1 \subset E_2 \subset X$,

(c) $\mu^*(\cup_n E_n) \leq \sum_n \mu^*(E_n)$ whenever $(E_n)_{n \geq 1} \subset X$.

Both (a) and (b) are clear. In proving (c), we may obviously assume that $\mu^*(E_n) < \infty$ for each n. Given $\epsilon > 0$, we can then choose open sets $(U_n)_{n \geq 1}$ such that $E_n \subset U_n$ and $\mu^*(U_n) < \mu^*(E_n) + \epsilon/2^n$ for each n. Put $U = \cup_{n \geq 1} U_n$. Given ϕ with $\phi \prec U$, the sets $(U_n)_{n \geq 1}$ form an open cover of $\operatorname{supp} \phi$, which is compact, so there exists a finite subcover $U_1, \ldots, U_N$, say. By Lemma A.3.4 we can find $\psi_1, \ldots, \psi_N \in C_c(X)$ such that $\psi_n \prec U_n$ for each n and $\operatorname{supp} \phi \prec \sum_1^N \psi_n$. Then $\phi = \sum_1^N \phi \psi_n$ and $\phi \psi_n \prec U_n$ for each n, so

$$\Lambda(\phi) = \Lambda\left(\sum_1^N \phi \psi_n\right) = \sum_1^N \Lambda(\phi \psi_n) \leq \sum_1^N \mu^*(U_n) \leq \sum_1^\infty \mu^*(U_n).$$

As this holds for each ϕ with $\phi \prec U$, we conclude that $\mu^*(U) \leq \sum_1^\infty \mu^*(U_n)$. Hence

$$\mu^*\left(\bigcup_n E_n\right) \leq \mu^*(U) \leq \sum_n \mu^*(U_n) \leq \sum_n \mu^*(E_n) + \epsilon.$$

Finally, letting $\epsilon \to 0$, we obtain (c).

The next step is to show that open sets are μ^*-*measurable*, i.e. that if U is open in X and E is an arbitrary subset of X, then

(A.2) $$\mu^*(E) \geq \mu^*(E \cap U) + \mu^*(E \setminus U).$$

To prove this, let V be an open set containing E, let $\phi \prec (V \cap U)$, and let $\psi \prec (V \setminus \mathrm{supp}\,\phi)$. Then $\phi + \psi \prec V$, so

$$\mu^*(V) \geq \Lambda(\phi + \psi) = \Lambda(\phi) + \Lambda(\psi).$$

As this holds for each ψ with $\psi \prec (V \setminus \mathrm{supp}\,\phi)$, we get

$$\mu^*(V) \geq \Lambda(\phi) + \mu^*(V \setminus \mathrm{supp}\,\phi) \geq \Lambda(\phi) + \mu^*(V \setminus U).$$

As this holds for each ϕ with $\phi \prec (V \cap U)$, it follows that

$$\mu^*(V) \geq \mu^*(V \cap U) + \mu^*(V \setminus U) \geq \mu^*(E \cap U) + \mu^*(E \setminus U).$$

Finally, as this is true for each open set V containing E, we conclude that (A.2) holds.

We now apply Carathéodory's lemma, which says that the collection of all μ^*-measurable sets forms a σ-algebra, and that the restriction of μ^* to this σ-algebra is a measure. Since we have shown that open sets are μ^*-measurable, it follows that all Borel sets are μ^*-measurable, and that the restriction of μ^* to the Borel σ-algebra is a Borel measure, which we shall denote by μ.

This measure μ is a Radon measure. In fact, if K is a compact subset of X, then

(A.3) $$\mu(K) \leq \inf\{\Lambda(\phi) : K \prec \phi\}.$$

To see this, take $\phi \in C_c(X)$ with $K \prec \phi$. Given $\alpha \in (0,1)$, if we put $U = \{x : \phi(x) > \alpha\}$, then U is open, $K \subset U$, and $\psi \leq \psi\phi/\alpha$ for all $\psi \prec U$. Hence

$$\mu(K) \leq \mu(U) \leq \sup\{\Lambda(\psi) : \psi \prec U\} \leq \sup\{\Lambda(\psi\phi/\alpha) : \psi \prec U\} \leq \Lambda(\phi)/\alpha.$$

Letting $\alpha \to 1$ gives $\mu(K) \leq \Lambda(\phi)$, which proves (A.3).

A.3. RADON MEASURES

The final step is to show that μ satisfies (A.1). To this end, take $\phi \in C_c(X)$, set $K = \operatorname{supp}\phi$, and fix $\beta > 0$ so that $\phi(K) \subset (-\beta,\beta)$. Given $\epsilon > 0$, choose $\gamma_0 < \gamma_1 < \cdots < \gamma_N$ such that

$$\gamma_0 = -\beta, \quad \gamma_N = \beta, \quad \text{and} \quad \gamma_n - \gamma_{n-1} < \epsilon \quad (1 \le n \le N).$$

For each n, put

$$E_n = \{x \in K : \gamma_{n-1} \le \phi(x) < \gamma_n\}.$$

Then by definition of μ^*, we can find an open set U_n containing E_n such that $\mu(U_n) < \mu(E_n) + \epsilon/N$. Shrinking U_n if necessary, we can further suppose that $\phi < \gamma_n$ on U_n. By Lemma A.3.4 there exist $\psi_1, \ldots, \psi_N \in C_c(X)$ such that $\psi_n \prec U_n$ for each n and $K \prec \sum_1^N \psi_n$. Then $\phi = \sum_1^N \phi\psi_n$ and $\phi\psi_n \le \gamma_n \psi_n$, so

$$\Lambda(\phi) = \sum_1^N \Lambda(\phi\psi_n) \le \sum_1^N \gamma_n \Lambda(\psi_n).$$

Now $\psi_n \prec U_n$ for each n, so $\Lambda(\psi_n) \le \mu(U_n)$. Also, since $K \prec \sum_1^N \psi_n$, it follows from (A.3) that $\mu(K) \le \Lambda(\sum_1^N \psi_n)$. Hence

$$\begin{aligned}
\Lambda(\phi) &\le \sum_1^N (\gamma_n + \beta)\Lambda(\psi_n) - \beta\Lambda\left(\sum_1^N \psi_n\right) \\
&\le \sum_1^N (\gamma_n + \beta)\mu(U_n) - \beta\mu(K) \\
&\le \sum_1^N (\gamma_{n-1} + \epsilon + \beta)(\mu(E_n) + \epsilon/N) - \beta\mu(K) \\
&\le \sum_1^N \gamma_{n-1}\mu(E_n) + (\epsilon + \beta)\mu(K) + (\beta + \epsilon + \beta)\epsilon - \beta\mu(K) \\
&\le \int_X \phi\, d\mu + \epsilon(\mu(K) + 2\beta + \epsilon).
\end{aligned}$$

Letting $\epsilon \to 0$, we conclude that $\Lambda(\phi) \le \int_X \phi\, d\mu$. Repeating the argument with ϕ replaced by $-\phi$ gives $\Lambda(\phi) \ge \int_X \phi\, d\mu$. Hence (A.1) holds, and the proof is complete. Phew! $\square$

A.4 Weak*-Convergence

Let X be a compact metric space. We denote by $C(X)$ the space of continuous functions $\phi\colon X \to \mathbb{R}$, equipped with the usual sup-norm, and by $\mathcal{P}(X)$ the collection of all Borel probability measures on X.

Definition A.4.1 A sequence $(\mu_n)_{n\geq 1}$ in $\mathcal{P}(X)$ is *weak*-convergent* to $\mu \in \mathcal{P}(X)$, and we write $\mu_n \xrightarrow{w^*} \mu$, if

$$\int_X \phi\, d\mu_n \to \int_X \phi\, d\mu \quad \text{for each } \phi \in C(X).$$

Theorem A.4.2 *Let X be a compact metric space. Then every sequence $(\mu_n)_{n\geq 1}$ in $\mathcal{P}(X)$ has a subsequence which is weak*-convergent to some $\mu \in \mathcal{P}(X)$.*

To prove this, we first need a separability lemma.

Lemma A.4.3 *If X is a compact metric space then $C(X)$ is separable, i.e. it has a countable dense subset.*

Proof. Since X is compact metric, it has a countable base $(U_n)_{n\geq 1}$ of open sets. For each pair m, n such that $\overline{U}_m \cap \overline{U}_n = \emptyset$, define $\psi_{m,n} \in C(X)$ by

$$\psi_{m,n}(x) = \frac{\operatorname{dist}(x, \overline{U}_m)}{\operatorname{dist}(x, \overline{U}_m) + \operatorname{dist}(x, \overline{U}_n)} \quad (x \in X).$$

The countable family of functions $(\psi_{m,n})$ separates the points of X. Let P be the collection of all finite products of the $\psi_{m,n}$, together with the constant function 1, and let Q be the set of finite $\mathbb{Q}$-linear combinations of elements of P. Then Q is still countable, and $\overline{Q}$ is a closed subalgebra of $C(X)$ which separates points and contains the constants, so by the Stone–Weierstrass theorem $\overline{Q} = C(X)$. Hence Q is a countable dense subset of $C(X)$. $\square$

Proof of Theorem A.4.2. Let $\{\phi_1, \phi_2, \ldots\}$ be a countable dense subset of $C(X)$. The sequence $(\int \phi_1\, d\mu_n)_{n\geq 1}$ is bounded, so it has a convergent subsequence $(\int \phi_1\, d\mu_n)_{n\in Z_1}$, say. Then the sequence $(\int \phi_2\, d\mu_n)_{n\in Z_1}$ is bounded, so it has a convergent subsequence $(\int \phi_2\, d\mu_n)_{n\in Z_2}$. Continuing in this manner, we obtain a nested family $Z_1 \supset Z_2 \supset Z_3 \supset \cdots$ of infinite subsets of the natural numbers such that, for each $j \geq 1$, the sequence $(\int \phi_j\, d\mu_n)_{n\in Z_j}$ is convergent. Let n_1 be the first element of Z_1, let n_2 be the second element of Z_2, and so on. Then $n_1 < n_2 < n_3 < \cdots$ and

A.4. WEAK*-CONVERGENCE

$(\int \phi_j \, d\mu_{n_k})_{k \geq 1}$ is convergent for each ϕ_j. Since the (ϕ_j) form a dense subset of $C(X)$, it follows that $(\int \phi \, d\mu_{n_k})_{k \geq 1}$ is convergent for every $\phi \in C(X)$. Thus we can define $\Lambda \colon C(X) \to \mathbf{R}$ by

$$\Lambda(\phi) = \lim_{k \to \infty} \left(\int_X \phi \, d\mu_{n_k} \right) \quad (\phi \in C(X)).$$

Then Λ is a positive linear functional on $C(X)$, so by the Riesz representation theorem (Theorem A.3.2) there exists a Borel measure μ on X such that

$$\Lambda(\phi) = \int_X \phi \, d\mu \quad (\phi \in C(X)).$$

This μ is a probability measure since

$$\mu(X) = \Lambda(1) = \lim_{k \to \infty} \left(\int_X 1 \, d\mu_{n_k} \right) = 1.$$

Thus $\mu \in \mathcal{P}(X)$, and clearly $\mu_{n_k} \xrightarrow{w^*} \mu$ as $k \to \infty$. □

We conclude by applying this theorem to prove a useful result about the existence of lifts.

Theorem A.4.4 *Let X and Y be compact metric spaces, and let $T \colon X \to Y$ be a continuous surjection. Then, given $\nu \in \mathcal{P}(Y)$, there exists $\mu \in \mathcal{P}(X)$ such that $\mu T^{-1} = \nu$, so that*

$$\int_X \phi(T(x)) \, d\mu(x) = \int_Y \phi(y) \, d\nu(y) \quad (\phi \in C(Y)).$$

Proof. For each $n \geq 1$, partition Y into finitely many Borel sets $(B_{n,j})$ of diameter less than $1/n$, and for each j pick an element $y_{n,j} \in B_{n,j}$. If we define

$$\nu_n = \sum_j \nu(B_{n,j}) \delta_{y_{n,j}},$$

where δ_y denotes the unit mass at y, then $\nu_n \in \mathcal{P}(Y)$ for all n. Also, given $\phi \in C(Y)$, the uniform continuity of ϕ implies that

$$\left| \int_Y \phi \, d\nu_n - \int_Y \phi \, d\nu \right| \leq \sum_j \nu(B_{n,j}) \sup_{y \in B_{n,j}} |\phi(y) - \phi(y_{n,j})| \to 0 \quad \text{as } n \to \infty,$$

and therefore $\nu_n \xrightarrow{w^*} \nu$ in $\mathcal{P}(Y)$.

Now as T is surjective, for each n, j there exists $x_{n,j} \in X$ such that $T(x_{n,j}) = y_{n,j}$. If we define

$$\mu_n = \sum_j \nu(B_{n,j}) \delta_{x_{n,j}},$$

then $\mu_n \in \mathcal{P}(X)$ and $\mu_n T^{-1} = \nu_n$. By Theorem A.4.2 there is a subsequence (μ_{n_k}) which is weak*-convergent to some $\mu \in \mathcal{P}(X)$. As T is continuous, it then follows that $\mu_{n_k} T^{-1} \stackrel{w^*}{\to} \mu T^{-1}$. Since also $\mu_n T^{-1} = \nu_n \stackrel{w^*}{\to} \nu$, we deduce that $\mu T^{-1} = \nu$, as desired. □

Bibliography

[1] L. V. Ahlfors and A. Beurling, 'Conformal invariants and function-theoretic null-sets', *Acta Math.* **83** (1950), 101–129.

[2] H. Alexander, 'Projections of polynomial hulls', *J. Funct. Anal.* **13** (1973), 13–19.

[3] J. M. Anderson and A. Baernstein, 'The size of a set on which a meromorphic function is large', *Proc. London Math. Soc. (3)* **36** (1978), 518–539.

[4] K. Astala, 'Area distortion of quasiconformal mappings', preprint, 1993.

[5] B. Aupetit, *Propriétés Spectrales des Algèbres de Banach,* Lecture Notes in Mathematics **735**, Springer-Verlag, Berlin, 1979.

[6] B. Aupetit, *A Primer on Spectral Theory,* Universitext, Springer-Verlag, New York, 1991.

[7] S. Axler, P. Bourdon and W. Ramey, *Harmonic Function Theory,* Graduate Texts in Mathematics **137**, Springer-Verlag, New York, 1992.

[8] T. Bagby and P. M. Gauthier, 'Approximation by harmonic functions on closed subsets of Riemann surfaces', *J. Analyse Math.* **51** (1988), 259–284.

[9] T. Bagby and N. Levenberg, 'Bernstein theorems', *New Zealand J. Math.* **22** (1993), 1–20.

[10] T. F. Banchoff and W. F. Pohl, 'A generalization of the isoperimetric inequality', *J. Differential Geom.* **6** (1971), 175–192.

[11] A. F Beardon, 'Integral means of subharmonic functions', *Math. Proc. Cambridge Philos. Soc.* **69** (1971), 151–152.

[12] A. F. Beardon, *A Primer on Riemann Surfaces*, London Mathematical Society Lecture Note Series **78**, Cambridge University Press, Cambridge, 1984.

[13] A. F. Beardon, *Iteration of Rational Functions*, Graduate Texts in Mathematics **132**, Springer-Verlag, New York, 1991.

[14] C. A. Berenstein and R. Gay, *Complex Variables*, Graduate Texts in Mathematics **125**, Springer-Verlag, New York, 1991.

[15] J. Bergh and J. Löfstrom, *Interpolation Spaces: An Introduction*, Grundlehren der Mathematischen Wissenschaften **223**, Springer-Verlag, Berlin, 1976.

[16] R. P. Boas, *Entire Functions*, Academic Press, New York, 1954.

[17] F. F. Bonsall and J. Duncan, *Complete Normed Algebras*, Ergebnisse der Mathematik und ihrer Grenzgebiete **80**, Springer-Verlag, New York, 1973.

[18] M. Brelot, *On Topologies and Boundaries in Potential Theory*, Lecture Notes in Mathematics **175**, Springer-Verlag, Berlin, 1971.

[19] H. Brolin, 'Invariant sets under iteration of rational functions', *Ark. Mat.* **6** (1965), 103–144

[20] D. M. Burns and S. G. Krantz, 'Rigidity of holomorphic mappings and a new Schwarz lemma at the boundary', *J. Amer. Math. Soc.* **7** (1994), 661–676.

[21] T. Carleman, 'Zur Theorie der Minimalflächen', *Math. Z.* **9** (1921), 154–160.

[22] L. Carleson, *Selected Problems on Exceptional Sets*, reprinted in Wadsworth Mathematical Series, Wadsworth, Belmont CA, 1983.

[23] L. Carleson and T. W. Gamelin, *Complex Dynamics*, Universitext, Springer-Verlag, New York, 1993.

[24] J. Deny, 'Sur les infinis d'un potentiel', *C. R. Acad. Sci. Paris Sér. I Math.* **224** (1947), 524–525.

[25] S. Dineen, *The Schwarz Lemma*, Oxford Mathematical Monographs, Clarendon Press, Oxford, 1989.

[26] A. Eremenko, G. Levin and M. Sodin, 'On the distribution of zeros of a Ruelle zeta-function', preprint, 1992.

[27] A. Eremenko and J. Lewis, 'Uniform limits of certain A-harmonic functions with applications to quasiregular mappings', *Ann. Acad. Sci. Fenn. Ser. A I Math.* **16** (1991), 361–375.

[28] J. Esterle, 'A complex-variable proof of the Wiener Tauberian Theorem', *Ann. Inst. Fourier (Grenoble)* **30** (1980), 91–96.

[29] K. J. Falconer, *Fractal Geometry : Mathematical Foundations and Applications*, Wiley, New York, 1990.

[30] T. W. Gamelin, *Uniform Algebras*, 2nd edition, Chelsea, New York, 1984.

[31] V. L. Garber, 'On the iteration of rational functions', *Math. Proc. Cambridge Philos. Soc.* **84** (1978), 497–505.

[32] J. B. Garnett, *Applications of Harmonic Measure*, University of Arkansas Lecture Notes in the Mathematical Sciences **8**, Wiley, New York, 1986.

[33] W. K. Hayman, 'Some applications of the transfinite diameter to the theory of functions', *J. Analyse Math.* **1** (1951), 135–154.

[34] W. K. Hayman and P. B. Kennedy, *Subharmonic Functions*, volume 1, London Mathematical Society Monographs **9**, Academic Press, London, 1976.

[35] W. K. Hayman, *Subharmonic Functions*, volume 2, London Mathematical Society Monographs **20**, Academic Press, London, 1989.

[36] L. L. Helms, *Introduction to Potential Theory*, reprinted by Robert E. Krieger, Huntington, New York, 1975.

[37] E. Hille and R. S. Phillips, *Functional Analysis and Semigroups*, A.M.S. Colloquium Publications **31**, American Mathematical Society, Providence RI, 1957.

[38] M. Klimek, *Pluripotential Theory*, London Mathematical Society Monographs New Series **6**, Clarendon Press, Oxford, 1991.

[39] S. G. Krantz, *Function Theory of Several Complex Variables*, 2nd edition, Wadsworth & Brooks/Cole, Pacific Grove CA, 1992.

[40] N. S. Landkof, *Foundations of Modern Potential Theory*, Grundlehren der Mathematischen Wissenschaften **180**, Springer-Verlag, Berlin, 1972.

[41] B. Ya. Levin, *Distribution of Zeros of Entire Functions,* Translations of Mathematical Monographs **5**, American Mathematical Society, Providence RI, 1964.

[42] J. L. Lewis, 'Picard's theorem and Rickman's theorem by way of Harnack's inequality', *Proc. Amer. Math. Soc.,* **122** (1994), 199–206.

[43] R. Mañé, P. Sad and D. Sullivan, 'On the dynamics of rational maps', *Ann. Sci. Ecole Norm. Sup. (4)* **16** (1983), 193–217.

[44] R. Nevanlinna, *Analytic Functions,* Grundlehren der Mathematischen Wissenschaften **162**, Springer-Verlag, Berlin, 1970.

[45] W. Parry and M. Pollicott, *Zeta Functions and the Periodic Orbit Structure of Hyperbolic Dynamics,* Astérisque **187–188**, Société Mathématique de France, 1990.

[46] J. R. Partington, 'Growth conditions on subharmonic functions and resolvents of operators', *Math. Proc. Cambridge Philos. Soc.* **97** (1985), 321–324.

[47] H. O. Peitgen and P. H. Richter, *The Beauty of Fractals: Images of Complex Dynamical Systems,* Springer-Verlag, Berlin, 1986.

[48] G. Pólya and G. Szegö, *Isoperimetric Inequalities in Mathematical Physics,* Princeton University Press, Princeton, 1951.

[49] Ch. Pommerenke, 'Über die Kapazität der Summe von Kontinuen', *Math. Ann.* **139** (1959), 127–132.

[50] Ch. Pommerenke, 'Über die Faberschen Polynome schlichter Funktionen', *Math. Z.* **85** (1964), 197–208.

[51] Ch. Pommerenke, *Boundary Behaviour of Conformal Maps,* Grundlehren der Mathematischen Wissenschaften **299**, Springer-Verlag, Berlin, 1992.

[52] F. Przytycki, M. Urbanski and A. Zdunik, 'Harmonic, Gibbs and Hausdorff measures on repellers for holomorphic maps, I', *Ann. of Math. (2)* **130** (1989), 1–40.

[53] T. Radó, *Subharmonic Functions,* Ergebnisse der Mathematik und ihrer Grenzgebiete, reprinted by Chelsea, New York, 1949.

[54] T. J. Ransford, 'A short proof of Johnson's uniqueness-of-norm theorem', *Bull. London Math. Soc.* **21** (1989), 487–488.

BIBLIOGRAPHY

[55] T. J. Ransford, 'A symmetrization inequality for plurisubharmonic functions', *Math. Ann.* **295** (1993), 191–200.

[56] T. J. Ransford, 'Variation of Hausdorff dimension of Julia sets', *Ergodic Theory and Dynamical Systems* **13** (1993), 167–174.

[57] W. Rudin, *Real and Complex Analysis*, 3rd edition, McGraw-Hill, New York, 1987.

[58] W. Rudin, *Functional Analysis*, 2nd edition, McGraw-Hill, New York, 1991.

[59] D. Ruelle, 'Repellers for real analytic maps', *Ergodic Theory and Dynamical Systems* **2** (1982), 99–107.

[60] M. Shishikura, 'The Hausdorff dimension of the boundary of the Mandelbrot set and Julia sets', S.U.N.Y., Stony Brook, Institute for Mathematical Sciences, Preprint 1991/7.

[61] A. M. Sinclair, *Continuous Semigroups in Banach Algebras*, London Mathematical Society Lecture Note Series **63**, Cambridge University Press, Cambridge, 1982.

[62] N. Steinmetz, *Rational Iteration: Complex Analytic Dynamical Systems*, de Gruyter Studies in Mathematics **16**, de Gruyter, Berlin, 1993.

[63] E. L. Stout, *The Theory of Uniform Algebras*, Bogden & Quigley, New York, 1971.

[64] M. Tsuji, *Potential Theory in Modern Function Theory*, 2nd edition, Chelsea, New York, 1975.

[65] P. Walters, *An Introduction to Ergodic Theory*, Graduate Texts in Mathematics **79**, Springer-Verlag, New York, 1982.

[66] J. Wermer, *Potential Theory*, 2nd edition, Lecture Notes in Mathematics **408**, Springer-Verlag, Berlin, 1974.

[67] J. Wermer, *Banach Algebras and Several Complex Variables*, 2nd edition, Graduate Texts in Mathematics **35**, Springer-Verlag, New York, 1976.

Index

accessible point, 113
Ahlfors–Beurling inequality, 141, 160
analytic multifunction, 207
approximation by
    continuous functions, 26
    harmonic functions, 176
    polynomials, 154, 169, 173, 176
    semicontinuous functions, 210
    smooth functions, 50, 72
    trigonometric polynomials, 12
asymptotic value, 102
    of holomorphic function, 104, 106
    of subharmonic function, 102
attracting basin, 190
    Green's function of, 192
    harmonic measure of, 194
    regularity of, 193
Aupetit, Bernard, 206

Banach algebra, 176
    commutative, 181
    radical, 186
    semisimple, 179
barrier, 88, 95, 126
Bernstein's lemma, 156
Bernstein–Walsh theorem, 170, 206
    converse to, 171
Beurling–Nevanlinna theorem, 120
    see also Ahlfors–Beurling
Bôcher's theorem, 83
Borel–Carathéodory inequality, 21, 23, 115
Borel measure, 209

Bouligand's lemma, 89
boundaries, convention about, 1, 6, 29
boundary, 181
    Shilov, 181
bounded approximate identity, 186
Bowen–Ruelle–Manning formula, 199
Brelot–Cartan theorem, 62, 65

Cantor set, 142
    capacity of, 143, 158
    Hausdorff dimension of, 198
    non-thinness of, 152
capacitable, 160
capacity, 127
    behaviour with respect to
        complements, 131
        decreasing sequences, 128
        increasing sequences, 128
        unions, 130
    estimate via
        arc-length, 138
        area, 141, 146
        diameter, 138, 140
        extremal length, 160
        length, 138
        moment of inertia, 146
        n-th diameter, 153, 168
        perimeter, 206
    of particular sets, 135
        Cantor set, 143, 158
        circular arc, 137
        connected set, 134, 138
        convex set, 206
        disc, 133

ellipse, 137
interval, 134, 136
Julia set, 191
Minkowski sum, 169, 206
rectifiable curve, 138
two intervals, 136
outer, 160
$\frac{1}{4}$-estimates for, 138
relation to Green's function, 132
relation to polynomials, 134, 154, 166
table of examples, 135, 160
via conformal mapping, 134
Carathéodory's lemma, 214
see also Borel–Carathéodory, Vitali–Carathéodory
Carleman–Milloux problem, 105, 120
Cartan, see Brelot–Cartan
character, 181
Chebyshev polynomial, 155
Choquet's theorem, 129, 144, 152, 160
closures, convention about, 1, 6
Cole, Brian, 160
commutative Banach algebra, 181
completely invariant, 190
conjugate indices, 161
conjugate of polynomial, 191
connected set,
capacity of, 134, 138
non-thinness of, 79
continuity principle, 54
convex function, 43
analogy with subharmonic function, 27
of subharmonic function, 43
convolution, 48, 165
distributional, 74
critical point, 198

decreasing sequence
of subharmonic functions, 37, 50, 78
Deny's theorem, 65, 83

diameter,
n-th, 152, 158, 168
of Julia set, 192
transfinite, 153
Dirichlet problem, 7, 85, 176
existence of solution,
on disc, 8
on regular domain, 91
to generalized Dirichlet problem, 95
non-existence of solution, 85, 88
uniqueness of solution, 7
disc, 1
capacity of, 133
equilibrium measure of, 75
n-th diameter of, 158
distribution theory, 74
domain, 1
dynamical interpretation,
of equilibrium measure, 195
of Green's function, 193
dynamics, 126, 190

energy, 55
entropy, 199
equilibrium measure, 58, 93, 159
of disc, 75
of Julia set, 195, 199
relation to harmonic measure, 105
uniqueness of, 75
ergodic theorem, 195
Evans' theorem, 66, 156

Fekete n-tuple, 152, 158, 159, 171
Fekete polynomial, 155
Fekete–Szegö theorem, 153
fine topology, 83
Fourier coefficients, 165
Fourier transform, 189
Frostman's theorem, 59
fundamental theorem of potential theory, 61

Gauss semigroup, 190

INDEX

Gelfand representation, 181
Gelfand transform, 181
gluing theorem, 37
Green's function, 106
    boundary values of, 111
    existence and uniqueness of, 106, 115
    integral formula for, 110
    joint continuity of, 115
    of attracting basin, 192
    positivity of, 107
    relation to capacity, 132
    subordination principle for, 107
        converse to, 111
    symmetry of, 110
    table of examples, 109

harmonic function, 3, 22
    approximation by, 176
    as real part of holomorphic function, 3, 5
    of holomorphic function, 5
    on Riemann sphere, 5, 22
    positive, 13
harmonic majorant, 118, 119, 124
    least, 118
harmonic measure, 96
    characterization of, 99
    estimation of, 105, 123, 126
    existence and uniqueness of, 96, 105
    for attracting basin, 194
    for half-plane, 104, 106
    for simply connected domain, 123, 125
    relation to equilibrium measure, 105
    subordination principle for, 101
    table of examples, 100
    zero, 100, 105, 146
Harnack distance, 14
    as a metric, 16, 21, 200
    subordination principle for, 15
Harnack's inequality,
    for discs, 13
    for general domains, 14
Harnack's theorem, 16
Hartogs' lemma, 83
Hausdorff dimension, 57, 198
Hausdorff measure, 57, 82, 198
Hausdorff–Young theorem, 165
Hilbert lemniscate theorem, 158
Hölder's inequality, 161
holomorphic family, 198
    Julia set of, 199, 207
holomorphic function,
    harmonic function of, 5
    real part of, 3
    subharmonic function of, 36, 39, 50
    zeros of, 41, 52, 76, 119, 124
holomorphic motion, 201
holomorphic semigroup, 186
    growth in radical Banach algebra, 186, 207
homogeneous polynomial, 166, 169
Hopf lemma, 51
hyperbolic, 198

identity principle,
    for harmonic functions, 6, 7
    for subharmonic functions, 50
increasing sequence,
    of harmonic functions, 16
    of subharmonic functions, 37, 62
integrability of subharmonic function, 39
integral means of subharmonic function, 46, 78
interpolation, 162, 206, 207
irregular point, 88
    criterion for, 92
    —s form polar set, 94
isoperimetric inequality, 141, 146, 160

Jensen's inequality, 43
    see also Poisson–Jensen
Johnson's theorem, 179, 206

Julia set, 190
  area of, 192, 198
  capacity of, 191
  diameter of, 192
  examples of, 191
  Hausdorff dimension of, 198
  of holomorphic family, 199, 207
  non-thinness of, 193

Keldysh's theorem, 173, 206
Kellogg's theorem, 94
Koebe's one-quarter theorem, 140, 145

L^p-norm, 161
L^p-space, 161
Laplace's equation, 3, 22
Laplacian, 1, 27, 36
  fundamental solution of, 74, 106
  generalized, 72, 74
Lebesgue measure, 39
  zero, 41, 56, 57
  see also Walsh–Lebesgue
Lewis, John, 17
lift, 217
lim sup of subharmonic functions, 64
Lindelöf's theorem, 104
  see also Phragmén–Lindelöf
Liouville theorem,
  for harmonic functions, 13, 21
  for subharmonic functions, 31, 47
  extended, 70
local uniform convergence, 11, 12, 16, 22, 64
logarithm,
  criterion for subharmonic, 44, 52
  existence of holomorphic, 4
lower semicontinuous function, 25

Mandelbrot set, 203
  main cardioid of, 203

Mañé, Sad and Sullivan, 200, 204
Manning, see Bowen–Ruelle–Manning
maximum principle,
  for harmonic functions, 6
  for holomorphic functions, converse to, 184
  for subharmonic functions, 29
  extended, 70

mean-value property, 5, 10
  converse to, 10
  local, 10, 11
Mergelyan's theorem, 172, 176
Milloux, see Carleman–Milloux
minimum principle, 55
Minkowski sum, 169, 206
minus infinity, 41, 42, 65
Montel's theorem, 22
μ^*-measurable, 214

nearly everywhere (n.e.), 56
  implies almost everywhere, 57
Nevanlinna, see Beurling–Nevanlinna

outer measure, 213

partition of unity, 212
Perron function, 86
  relation with Poisson integral, 96, 97
Perron method, 86
Phragmén–Lindelöf principle, 30, 51, 51
  for sector, 32
  for strip, 31
Picard's theorem, 17, 23
PL-functions, 52
pluriharmonic function, 22
pluripotential theory, 22
plurisubharmonic function, 206
Poisson integral, 8
  generalized, 96
  relation with Perron function, 97

INDEX

Poisson integral formula, 10, 35
Poisson–Jensen formula, 117
    classical, 118
Poisson kernel, 8, 9, 12
Poisson modification, 86
polar set, 56, 160
    conformal invariance of, 70
    connected complement, 68
    harmonic measure zero, 100
    Lebesgue measure zero, 56, 57
    non-Borel, 58, 83, 144, 160
    thinness of, 79, 82
    totally disconnected, 69, 81, 144
    uncountable, 57, 65, 143
polynomial,
    approximation by, 154, 169, 173, 176
    attracting basin of, 190
    Chebyshev, 155
    Fekete, 155
    homogeneous, 166, 169
    hyperbolic, 198
    Julia set of, 190
    quadratic, 203
    trigonometric, 12
polynomially convex, 157
potential, 53

quadratic polynomial, 203

radical, 179
radical Banach algebra, 186
Radó–Stout theorem, 69, 71
Radon measure, 72, 211
reflection principle, 11, 12, 116
regular domain, 88
    attracting basin, 193
    conformal invariance of, 116
    simply connected domain, 92
regular measure, 210
regular point, 88
    criterion for, 92, 111
    relation to thinness, 93
removable singularity theorem, 67
Riemann mapping theorem, 113
    extension of mapping function to closure, 114, 116, 126
Riemann sphere, 1, 5, 22, 36
Riemann surface, 22
Riesz decomposition theorem, 76
Riesz representation theorem, 72, 211
Riesz–Thorin interpolation theorem, 162, 206
Rudin's theorem, 184, 206
Ruelle, *see* Bowen–Ruelle–Manning
Runge's theorem, 169, 171
    quantitative version of, 170

Sad, *see* Mañé, Sad and Sullivan
Schwarz's lemma, 126
    at boundary, 51
semisimple Banach algebra, 179
Shilov boundary, 181
simply connected domain,
    conformal mapping to disc, 113
    harmonic measure on, 123, 125
    necessity for, 5, 204
    regularity of, 92
smoothing theorem, 49
spectral radius, 177, 178
spectral radius formula, 177
spectrum, 177, 207
Stout, *see* Radó–Stout
subharmonic function, 28
    convex function of, 43
    discontinuous, 29, 41, 78
    of holomorphic function, 36, 39, 50
    on Riemann sphere, 36
    radial, 45
    set where $-\infty$, 41, 42, 65
    smooth, 36
    smooth approximation of, 50
submean inequality,
    global, 35
    local, 28, 51
subordination principle,
    for Green's functions, 107
    converse to, 111

for harmonic measure, 101
for Harnack distance, 15
subsequence, convergent,
   of harmonic functions, 16, 125
   of measures, 58, 125, 216
   of subharmonic functions, 126
Sullivan, *see* Mañé, Sad and Sullivan
sup-norm, 72, 154
superharmonic function, 28
support of a measure, 53, 209
supremum of subharmonic functions, 38, 62
symmetrization inequality, 166
symmetry theorem, 110
Szegö, *see* Fekete–Szegö

table of examples,
   of capacities, 135, 160
   of Green's functions, 109
   of harmonic measure, 100
thin set, 79
   necessary condition for, 150
   non-thinness of
      Cantor set, 152
      connected set, 79
      Julia set, 193
   set at n.e. point of itself, 81
   relation to regularity, 93, 176
   thinness of polar set, 79, 82
   Wiener's criterion for, 146, 152
Thorin, 206
   see also Riesz–Thorin
three-lines theorem, 32
transfinite diameter, 153
trigonometric polynomial, 12
two-constant theorem, 101

uniqueness-of-norm theorem, 179
upper semicontinuous function, 25
   approximation by continuous functions, 26
   compactness theorem for, 25
   continuity at a dense G_δ set, 27

upper semicontinuous regularization, 51, 62

Vesentini's theorem, 178, 206, 207
Vitali–Carathéodory theorem, 98, 210

Walsh–Lebesgue theorem, 173, 176, 185
   see also Bernstein–Walsh
weak*-convergent, 58, 216
   subsequence, 58, 125, 216
Wermer's theorem, 182, 207
Weyl's lemma, 76
Wiener's criterion, 146, 152
Wiener's tauberian theorem, 189

Young's inequality, 165
   see also Hausdorff–Young

Glossary of Notation

Symbol	Meaning	Page
$\mathbb{C}$	complex plane	1
$\mathbb{C}_\infty$	Riemann sphere	1
$\overline{S}$	closure of S relative to $\mathbb{C}_\infty$	1
∂S	boundary of S relative to $\mathbb{C}_\infty$	1
$\Delta(w, \rho)$	open disc, centre w, radius ρ	1
$\overline{\Delta}(w, \rho)$	closed disc, centre w, radius ρ	1
Δf	Laplacian of f	1
$P(z, \zeta)$	Poisson kernel	8
$P_\Delta \phi$	Poisson integral of ϕ on Δ	8
τ_D	Harnack distance on D	14
1_S	characteristic function of S	26
dA	two-dimensional Lebesgue measure	39
$M_u(r)$	maximum of $u(z)$ on circle $|z| = r$	46
$C_u(r)$	average of $u(z)$ on circle $|z| = r$	46
$B_u(r)$	average of $u(z)$ on disc $|z| \le r$	46
U_r	set of points of U at distance $> r$ from boundary	48
$u * \phi$	convolution of u and ϕ	48
$\mathrm{supp}\,\mu$	support of μ	53
p_μ	potential of μ	53
$I(\mu)$	energy of μ	55
n.e.	nearly everywhere	56
$\mathcal{P}(K)$	the Borel probability measures on K	58
u^*	upper semicontinuous regularization of u	62
$C_c^\infty(D)$	the C^∞ functions on D of compact support	71
Δu	generalized Laplacian of u	72
$C_c(D)$	the continuous functions on D of compact support	72
$\|\phi\|_\infty$	sup-norm of ϕ	72
$\partial_e K$	exterior boundary of K	75

GLOSSARY OF NOTATION

Symbol	Meaning	Page
$H_D\phi$	Perron function on D associated to ϕ	86
$\mathcal{B}(\partial D)$	the Borel subsets of ∂D	96
ω_D	harmonic measure for D	96
$P_D\phi$	generalized Poisson integral of ϕ on D	96
g_D	Green's function for D	106
$c(E)$	capacity of E	127
$C(\mathbf{s})$	generalized Cantor set corresponding to sequence $\mathbf{s}$	143
$\delta_n(K)$	n-th diameter of K	152
$\|q\|_K$	sup-norm of q on K	154
$m_n(K)$	minimum $\|q\|_K$ for monic polynomial q of degree n	155
$\|f\|_p$	L^p-norm of f	161
$L^p(\mu)$	L^p-space of μ	161
$f * g$	convolution of f and g	165
$d_n(f, K)$	minimum $\|f - p\|_K$ for polynomial p of degree $\leq n$	170
$A^\#$	Banach algebra A with identity adjoined	176
$\sigma(a)$	spectrum of a	177
$\rho(a)$	spectral radius of a	177
$\mathrm{rad}(A)$	radical of A	179
Φ_A	character space of A	181
$\widehat{f}$	Gelfand transform of f	181
∂_A	Shilov boundary of A	181
$\rho_{f,g}$	maximum of $|\widehat{f}|$ on $\widehat{g}$-fibre	182
F_∞	attracting basin of ∞	190
J	Julia set	190
$m_\alpha(S)$	α-dimensional Hausdorff measure of S	198
$\dim(S)$	Hausdorff dimension of S	198
$e_\mu(q)$	entropy of q with respect to μ	199
M	Mandelbrot set	203
W	main cardioid of Mandelbrot set	203
$\mathrm{supp}\,\mu$	support of μ	209
1_B	characteristic function of B	211
$C_c(X)$	the continuous functions on X of compact support	211
$K \prec \phi$	$\phi \in C_c(X)$, $0 \leq \phi \leq 1$, $\phi = 1$ on K	212
$\phi \prec U$	$\phi \in C_c(X)$, $0 \leq \phi \leq 1$, $\mathrm{supp}\,\phi \subset U$	212
$C(X)$	the continuous functions on X	216
$\mathcal{P}(X)$	the Borel probability measures on X	216
$\mu_n \stackrel{w^*}{\to} \mu$	μ_n is weak*-convergent to μ	216